STATISTICAL DATA ANALYSIS FOR OCEAN AND ATMOSPHERIC SCIENCES

STATISTICAL DATA ANALYSIS FOR OCEAN AND ATMOSPHERIC SCIENCES

H. Jean Thiébaux
Department of Meteorology
Pennsylvania State University
*University Park, Pennsylvania**

**Current Address:* *Fort*
Washington, Maryland

ACADEMIC PRESS
San Diego   New York   Boston   London   Sydney   Tokyo   Toronto

*"MINITAB" is a registered trademark of Minitab Inc., 3081 Enterprise Drive, State College, PA 16801-3008; telephone: (814) 238-3280.

This book is printed on acid-free paper. ∞

Academic Press, Inc.
A Division of Harcourt Brace & Company
525 B Street, Suite 1900, San Diego, California 92101-4495

United Kingdom Edition published by
Academic Press Limited
24-28 Oval Road, London NW1 7DX

Library of Congress Cataloging-in-Publication Data

Thiébaux, H. J.
    Statistical Data analysis for ocean and atmospheric sciences / by H. Jean Thiébaux.
        p. cm.
    Includes bibliographical references and index.
    ISBN 0-12-686925-1 ISBN 0-12-686926-X (Diskette)
    1. Oceanography--Statistical methods. 2. Heterology--Statistical methods. I. Title.
    GC10.4.S7T48   1994
    551.46'0072--dc20   94-10463
                                                                  CIP

PRINTED IN THE UNITED STATES OF AMERICA
94 95 96 97 98 99 QW 9 8 7 6 5 4 3 2 1

To the memory of my grandfather,
Frank Loxley Griffin,
and his gift for teaching mathematics to
students of the arts and sciences.

CONTENTS

2

DATA AND DATA MANAGEMENT
"what we have to go on" or "accumulated records
of observations and their expeditious reorganization" 15

3

DESCRIPTIVE STATISTICS
"first impressions" or "sketching features of observed systems
with data" 33

4

THE FOUNDATIONS OF INFERENCE
"probability models as descriptions
of research outcomes" 53

5

STOCHASTIC VARIABLES AND THE IDENTIFICATION OF THEIR DISTRIBUTIONS
"distilling uncertainty"   73

6

THE EXPONENTIAL AND UNIFORM DISTRIBUTIONS
"describing uncertainty in time and space"   95

7

THE NORMAL DISTRIBUTIONS
"good approximations for many composite variables"   111

8

ANALYZING VARIABILITY
**"establishing differences between means
and between variances" 143**

9

TESTING HYPOTHESES
**"dealing with the generic critic while establishing
powerful support for new ideas" 173**

10

LINEAR REGRESSION
"analyzing an influence network" 195

11

BOOTSTRAPPING
"scientific inference when none of the above apply"   225

PREFACE

Scientific knowledge of the oceans and atmosphere comes from interpretations of data—with all their attendant errors. The challenge to scientists is to filter truth from data. Sometimes the values of a data set are sequentially or spatially correlated; generally the data are gathered from diverse observing systems with inhomogeneous error properties; and always we lack "insider information" on the details of their statistical structure. The validity of deductions about the truth depends on the finesse with which data sets are handled, and this is the business of statistical data analysis.

Statistical Data Analysis for Ocean and Atmospheric Sciences is an uncommon introduction to methods of statistical analysis. It is a thinking person's guide to scientific inference when the data base provides all the evidence there is, and human reasoning abilities and computational software are the only available tools.

The text has been written as a guide for students through what may once have seemed to be a maze of descriptive methods, concepts, and inferential techniques. Our objective is to clarify the relationships among these descriptive methods, statistical concepts, and inferential techniques, as well as to provide an understanding of them in the contexts of scientific inference.

The book comes with a well-packed data disk and instructions on the use of universally available statistical software for exercising the definitions, concepts, and techniques as they are introduced. The data files and the statistical explanations in contexts of ocean and atmospheric sciences make the book unique and, I hope, of unique value to those who use it in their studies. The data files have been assembled from data sets of meteorology and oceanography. The exercises using these data with the prescribed computational software demonstrate the practicality and applicability of statistical analysis in these disciplines.

H. Jean Thiébaux

ACKNOWLEDGMENTS

The work on this book was begun with the support of the Visiting Professorships for Women Program of the National Science Foundation and the hospitality of the Pennsylvania State University. It gained breadth and focus through the endurance and honesty of my students. It was finished with the encouragement of my friends and children. At times their vision was greater than mine and it carried me through to completion. I gratefully acknowledge all those who contributed to this task. May the end result be worthy of what they have given.

I

STATISTICAL ANALYSIS AND INFERENCE IN SCIENCE

"the art of reaching conclusions at the interface of theory and observation"

1.1 INTRODUCTION

This book is about the making of scientific inferences: the joining of theory and fresh observations. It is about coalescing new information and a theoretical framework, when what we are able to observe is only a portion of the system of interest and our records of observations include extraneous innovations.

You are students of science and you have chosen to focus on understanding the mechanisms of circulations and energy transport within our oceans and atmosphere. What we know of these mechanisms is continually being updated. As we gain more complete information from Earth observing systems and experiments with computerized models based on prior information, we reevaluate and revise our understanding. Our goal is to use new information to bring our characterizations of the dynamics of the oceans and atmosphere closer to truth—a process which requires sophisticated management of uncertainty.

No matter how well planned a field experiment has been, the resulting data set will not completely describe the system we have observed, nor will it be free from measurement inaccuracies and influences of unrepresentative small-scale phenomena. It is easier to imagine writing a perfect four-hour exam paper in geophysical fluid dynamics, than it is to imagine a field experiment being completed without a hitch, even if its location and time are as near as the roof of the building in the first strong wind of the semester! A scarf blowing over your eyes at a critical moment, a dropped thermometer, coffee spilled on a notebook entry, ..., each contributes unplanned uncertainty in the records of the experiment. And accidental causes for uncertainties are not limited to small scale experiments.[1] Despite flaws in execution and despite the inaccuracies of sensing and recording equipment, we analyze the data we have obtained and reach scientific conclusions. This book is about strategies for handling data sets, with all their baggage of incomplete records and susceptibility to errors, to arrive at valid inferences.

We include a thorough examination of the statistical foundations of these strategies, whose language, notation, and concepts are central to accurate interpretation of research results and to presentation of conclusions to colleagues and critics. When you have become familiar with the foundation stones of scientific inference, composed of statistical distributions and techniques for estimation and testing of hypotheses, you will see that they provide a formalism for scientific inference. Specifically, they permit us to juxtapose new experiences with Earth's fluid envelope, as they are recorded in data sets, on theoretical/hypothetical frameworks, and assess the levels of agreement. The validity of each assessment is dependent on the realism of the choice of stochastic representation for the collective elements of uncertainty in the data. Thus, understanding your options and their consequences for inferential validity is of the greatest importance to the computation, presentation, and defense of your hard won results. This text provides a guide to building a solid basis for your conclusions.

Requirements of a course in statistical techniques, as applied to inference in a scientific field, necessarily evolve with the evolution of available computational facilities, as well as with the focus of the scientific research which it serves. The remarkable developments in computer technology in the latter part of the 20th century have made it possible to deal efficiently and accurately with large data sets. From the first assemblage of data (frequently, from the point of initial observation), they are handled electronically. Observed values from bathythermographs, radiosonde stations, and satellite sensors are transmitted via satellite tele-

---

[1]As we know from reports by the press, even the most expensive and thoroughly planned experiments may reveal accidents in their preparation that flaw results in totally unexpected ways. An outstanding example that will long remain in the archives of science is the Hubble telescope which was sent to observe the solar system from outside Earth's atmosphere. The telescope had a major fault in its construction that was not detected until it was in Earth orbit.

connections to computer data bases. Data quality checking may be done automatically; and data can be used immediately or archived in computer-based storage for later use. The computation time limitations which once inhibited thorough analysis of data sets are no longer a factor. We are at liberty to include any data base relevant to our inquiry, virtually without concern for its size; and the statistical analysis through which we filter it is limited only by scientific insight. Whether the data of your own research project has been collected and tabulated by hand or transferred to a computer file by ARGOS, you will wish to use a computer for its analysis. Accordingly, in this book, all of our conceptual illustrations are made using available computer software. This is the data analytic technology of the present and future.

The conceptual framework and notation for the statistical methods presented in this book are derived from classical statistics. The contexts in which they are developed and which provide the illustrative examples have been selected from research reports appearing in recent literature of the ocean and atmospheric sciences. They have been chosen to demonstrate the relevancy and power of statistical inference to scientific inquiry and to facilitate in transferring the principals of stochastic modeling, estimation, and hypothesis testing, to other research situations. You will discover that we emphasize confirmation of the premises of standard statistical tests, wherever they are discussed. This is because most of these tests were developed in rather different research settings where the variables of the data sets were not drawn from spatially coherent media. Thus their premises may be inappropriate for inference in some geophysical contexts. If we find this is true, then we must make an extra investment in Monte Carlo simulation, in order to achieve inferential validity.

1.2 A FORMAT FOR STATISTICAL INFERENCE

Research begins with a question. We may seek a descriptive or explanatory result, or a measure of association of two or more phenomena. Whatever the specific nature of our objective in conducting a research project, we can describe our purpose as answering a scientific question. The statement of the scientific question is basic to the design of the research.

From the initiating question we turn our attention to determining what it will take to arrive at an answer. Specifically, we focus on practical aspects of the data that can be made available: their spatial/temporal characteristics and error structure, and the identification of appropriate analysis procedures. The three elements are closely linked with one another; and the process of reaching a final research design may take several iterations. Discovery of practical constraints on the collection and storage of information may lead to revision of the research question. This in turn will influence the choice of mechanism for focusing the information, in formulating answers or conclusions. In each inference example discussed in the text we will follow a logical format, by answering the following questions in order.

- What is the scientific objective: What research question do we seek to answer?
- What information can we bring to bear in answering this question; and what are the sources of uncertainty associated with this information?
- How will we use the information we obtain from observations to answer the research question, i.e., to arrive at scientific conclusions in the contexts of our theoretical framework and known sources of uncertainties?

This is the format of statistical inference. Coupled with actual data, the structure provides answers and measures of the confidence that can be associated with them. Not all the queries in this hierarchy have short answers. However, when you have learned and followed the logical structures of inference, from scientific question to scientific conclusions, you will be able to assign, report, and defend your conclusions *and* the level of confidence you associate with them.

The first query above requires a succinct formulation of the objective of the research. In practice, as you contemplate answers to the queries which come after it, you will find yourself reformulating the basic research question, most likely tightening it up, so that it is a more precise statement of a hypothesis in juxtaposition with the evidence that can be obtained to support the hypothesis. It is good to anticipate this reformulation and see it as an important part of scientific inference. In fact, every half hour you spend rethinking and restating these questions so that they are more clearly focused will undoubtedly save you hours of research time, if not days or weeks!

The answer to the second query in the format outline identifies variables which can be measured, or for which records exist, and whose values, taken together, may elicit research conclusions. Identifying specific variables which will be used is closely tied to restating the research question: framing the question as a question that is answerable in terms of observable variables. In addition, we must know how the variables will be observed and how the observations will be transmitted and recorded. This is in order to correctly characterize the uncertainty associated with the data we will use. If the data already exist in an archive, we will require a description of the error characteristics of the recorded values.

The answer to the final query in our outline must anticipate possible outcomes to the execution of the specific research strategy, because it sets up a decision-making algorithm. The algorithm is based on the relative likelihoods of outcomes, given the competing research hypotheses *and* the known characteristics of uncertainty in the data. This is the final step. Specification of the mechanisms for evaluating research hypotheses in light of the discriminating data which will be produced by the program, completes the research plan. When this is in place we are ready to objectively evaluate whatever outcome the research strategy produces.

The following are important notes on the process of research design.

(i) It is highly recommended that the decision-making algorithm be formulated prior to obtaining or examining any data.

(ii) The formulation of the decision-making algorithm permits us to quantitatively relate the distinctions we wish to be able to make to the amount of data required to make them. If we have not planned for sufficient data to achieve our objective, then we will need to revise the plan.

1.3 PICKING ANALYSIS TECHNIQUES AND SEEKING HELP

As a statistical consultant, I have responded to requests for assistance with analyzing data on numerous occasions. Most frequently the request has been some variation of: "This is a really simple question. I have this data and I can't figure out exactly how to analyze it, although I am sure it's straightforward. Will you help me with it"?

It is important to realize, as you approach this course and your future in reaching scientific conclusions from theory and observation, that statistical evaluations are generally not either simple or straightforward. Furthermore, it is generally unrealistic and unnecessary to expect that, working by yourself, you will be able to identify the most efficient inferential tools to bring your research to fruition. To request and participate in analytical collaboration with a statistical scientist is to venture on firm ground to confident scientific conclusions. A good statistician is a good partner to involve in your research from its inception.

The algorithms for estimating parameters of physical fields and making scientific decisions are logical constructions. In any specific research setting, these are derived directly from the statement of the scientific objective, coupled with descriptions of the observed variables and the mechanism for obtaining and reporting their observations. Because this is true, the point of departure for arriving at an appropriate analysis scheme is via:

(i) Statement of the scientific objective;

(ii) Description of the variables selected for observation and of how the data will be obtained, including description of the sources of uncertainty: noise and measurement errors in the data collection system;

(iii) Specification of the use of the data, with management of uncertainty, in answering the research questions imbedded in the scientific objective.

This highly structured point of departure is in bold contrast with waiting until you have assembled data to be thoroughly rigorous in stating your scientific objective. It may look like taking extra steps. However, it leads to relevant analysis and valid conclusions; and once you have tried it, you will own two important bits of knowledge. The first is that you have in hand all the information required by the format of statistical inference; and the second is that each element of information in (i)–(iii) is critical to deductions about a physical system when our experience of it contains uncertainties. Frequently, in fact, selection of the appropriate analysis technique can be arrived at by a series of restatements of (i)–(iii), as you might rephrase an explanation or description given to a colleague when you wish to be perfectly understood. Each iteration in the series brings in more detail as you recognize its relevancy, until you see that the pattern of your knowledge and assumptions about the system studied, in juxtaposition with what you are committed to learning about it, identifies an analysis algorithm. By this mechanism you will fully understand and then be able to defend the inferential process which led to your conclusions. In carrying through this process, do not hesitate to seek statistical partnership, for statistics is an art: *the art of persuading the universe to yield information about itself*.

Whatever the focus of your own immersion in the scientific process of discovery, techniques of statistical data analysis will play a role: in estimating characteristics of the system under observation and in evaluating the validity of an hypothesis in the face of accumulated data. You may choose to construct a statistical technique from your specific premises or adapt one from someone else's work. In each case it is wise to reason through the derivation of the technique, noting all the assumptions which have gone into its construction and comparing them with your known reality. If any of them fail the "truth test", the validity of the application of the technique is in question.

1.4 A FEW EXAMPLES

Table 1.1 presents a list of questions which have been posed for actual research projects. In this section we will use two of these to illustrate the process of restating questions as questions which can be answered with data that are or could be available. The remainder will be used in

TABLE 1.1   Research Questions to Be Addressed by a Research Plan

1. Would an automated cable TV tornado warning system in southeastern Texas be effective?
2. Can winter weather be accurately foretold, based on the previous fall season's weather?
3. Is the number of tornados generated by a hurricane related to the strength/intensity of the hurricane?
4. What effect does air pollution have on urban climate?
5. What are the magnitudes and frequencies of extreme rainfalls in the watershed of the Susquehanna River upstream of its junction with the Juniata River?
6. How does the salinity of the ocean vary with temperature and depth?
7. Are wind speed forecasts of the MOS/NGM (Model Output Statistical regressions using output of the Nested Grid Model) correlated with the discrepancies between observed minimum temperatures and MOS/NGM forecast temperatures?
8. What atmospheric conditions are best for using an electro-optic sensor (EOS)?

exercises, to give you practice in articulating a scientific objective in unambiguous terms.

Example 1

The first question has considerable practial significance to residents of southeastern Texas, as well as to the cable company which would provide the automated warning system. If it were effective, it would support protective measures saving lives and property. If it were not effective, there would be no reason for the cable company to invest resources in its provision. We note that the statement of the research question does not identify a clear path to an answer. Both "automated cable TV tornado warning system" (ACTVWS) and "effective" are not defined. We must be clear about just what the service would provide and we must define a measure of effectiveness. To be specific, let us say that ACTVWS would relay any tornado warning received from National Oceanic & Atmospheric Administration (NOAA) weather radio, for the broadcast area, with a tone and warning message overriding all cable channel programming. "Effective" clearly means alerting the public to an anticipated tornado, with sufficient lead time that people can take protective action. Having said this, we can see that we have just begged answers to three questions:

   (i) What information would be made available via ACTVWS that permits taking protective action; and at what time relative to an anticipated event, would this reach the public?
  (ii) What is "sufficient lead time"?
 (iii) How do we measure effectiveness in alerting the public?

If we can answer these questions, then we can phrase the research objective in terms which can be addressed unambiguously.

A tornado warning is issued by NOAA when a tornado has been sighted or when there are radar indications that there will be a tornado within a one- to two-county radius, within an hour. Accordingly, the information that could be made available via ACTVWS is that a tornado is expected to be imminent within a specific area. Because of the circumstances that trigger the issuance of a tornado warning by NOAA, there is no "lead time" per se. Protective action must be taken immediately to avert damage to life and property. Thus, to be effective, the cable company will need to maintain connection with NOAA weather radio, around the clock, and relay warnings as they come in, without delay. We will assume that the cable company has the technology to make the relay connection, and that the proposal is to use this technology. (Otherwise the research question is vacuous, since there is no lead time for a tornado warning with our current state of knowledge.)

Finally, we need to establish a measure of "effectiveness in alerting the public", in the context of a tornado warning. A reasonable measure would be the proportion of the population of the designated area who receive the relayed warning within, say, five minutes of its being issued by NOAA weather radio. Providing this answer will require an accurate survey of daily activity schedules *with respect to television viewing*, covering all ages and occupations, from which a composite picture can be constructed of the population within shouting range of a cable TV viewer. This is a probably not what came to mind when you first read question (i), on the list above. (At least, it was not what came to my mind.) However, now that we have arrived at this point, we see that we are getting closer to an appropriate, answerable question with:

> "What proportion of the population is within shouting distance of a cable television viewer at any given time?"

This will work for us, in the sense that we can obtain survey information from which we can approximate an answer. If we use an appropriately randomized survey, we can also put "assurance bounds" on the outcome.

Now that we have rephrased the research question, so that it is a specific question with a quantitative answer, we may realize that we wish to be more specific. For example, we might wish to rephrase it one final time as:

> "What are the proportions of urban and rural households, schools, and businesses which have at least one cable television viewer at any time?"

This question will need to be answered for different times of the day, and separately for regular work days, weekends, and holidays. Note that the present statement of the research question does not look at all like the original question. However, it *is* what we want to know; and it is possible to answer with a bit of survey research.

There is no doubt that this kind of rephrasing of the research question takes the romance out of it. One way of saying this is that it takes us from the sublime to the mundane. It forces us to articulate precisely what it is that we want to know and the quantitative measures by which we will arrive at an answer. This may take one rephrasing. More commonly, it will involve many. However, the process provides its own reward: When it is complete, the steps required to reach our objective will be outlined in terms of familiar variables which we know how to measure.

Example 2

The person who phrased the question:

"Can winter weather be foretold from the previous fall's weather?"

planned to use the most recent 50 years of the temperature record for State College, PA, to check out an old meteorological saying: "If we have a warmer than normal fall, expect a colder than normal winter." We can rephrase the initial question to match this objective more closely, as:

"Based on the most recent 50-years' record, can we conclude that a fall with above normal temperatures is more likely to be followed by a winter with below normal temperatures?"

This is the first big step toward clear definition of the research project. Next we must agree on what we will mean by "temperatures above (below) normal"; and, finally, we must establish a criterion for deciding whether the evidence supports the belief.

Traditionally, whether a season is "above normal" or "below normal" is determined by comparison of the daily average temperatures for the season with daily climatological values. With a 50-year temperature record a "50-year climatological value" for each day of the year is obtained by summing the 50 average temperature values for that day of the year and dividing by 50. That is, the climatological value is the arithmetic mean for the day of the year. We will exclude all the February 29 dates and create a data set with 365 values: each an average of 50 daily average temperatures. Those designated as "fall" will be the 91 values corresponding to September 21 through December 20. Those designated as "winter" will be the 90 values corresponding to December 21 through March 20. By convention, temperatures of a season of a specific year are said to be *above normal* if the number of days of the season for which the daily average temperatures exceed their climatological values, say N^+, is greater than the number of days for which the daily average temperatures are less than the corresponding climatological values, say N^-. Alternatively, using the value $(N^+ - N^-)$, we say that the season is warmer than normal if this difference is positive and colder if it is negative. Let X denote this difference for a summer season and Y for a winter season.

From a 50-year record, beginning with January 1 and ending with December 31, 50 years later, we have 49 pairs: a fall followed by a winter which is completely within the record. Thus we have corresponding values X_j and Y_j, for $j = 1, \ldots, 49$, comprising the evidence on which we will base our conclusion. At this point, we can only speak generally and say that we will wish to conclude that warmer than normal fall seasons are generally followed by colder than normal winters if we find that among the pairs $(X_1, Y_1), \ldots, (X_{49}, Y_{49})$, positive values of X are generally associated with negative values of Y. It will be easier to be specific about the meanings of "generally followed by" and "generally associated with", when we have discussed concepts of chance associations. We will return to this example later.

1.5 DATA SETS AND DATA MANAGEMENT

The exercises at the end of each chapter will use real data sets. These data sets are stored in ASCII format files, on the disk included with this book, and are described in Chapter 2. The data have been assembled and published with the text to demonstrate and encourage the use of computer software on moderately large sets of real data. The software we have recommended is designed to carry out all the data management and computational tasks associated with statistical data analysis, so that long-hand tabulations and the use of a calculator will not be required.

Chapter 2 also contains a description of the use of MINITAB for the selection of values within a data set according to a specific criteria, with some exercises to familiarize you with these techniques. For example, from a data set containing daily reports of a variable over a number of years, we may wish to consider the reports from only one or two of the years, or those for only a single month, or values which exceed one standard deviation of the mean. These selections, which would be difficult without efficient software, are simple maneuvers once you have had just a little practice with MINITAB or a comparable statistical software package. (If you have not yet had an introduction to these facilities, you are about to discover one of the wonders of current technology. You will be pleased when you have acquired this skill and relieved that its acquisition was so easy.)

1.6 FINAL TECHNICAL NOTE

As with all subcultures, different "subcultures of science" assign different meanings and notations to a few key words and concepts. It is virtually impossible to address an interface between disciplines in language and notation that honors all the conventions within the disciplines. You will find that to be true here, as your statistical expertise grows in relation to your own field. You cannot completely avoid the shocks of seeing some of your old standbys redefined and used in heretofore unfamiliar ways. We ask that you adopt a liberal attitude in this regard; and pay close attention to definitions as they come into the script. That will help to keep the progression of ideas in focus: from data management to inferential conclusions.

A good example of a contrast in definition and usage is provided by the concept of a vector. We will rarely use this in the sense of a directed force or velocity. Throughout the text, *vector* will be given the more general interpretation of a many $\times 1$ array of values. It will be used in contexts where maintaining the ordering or identification of the values is essential to data analysis objectives. Thus a vector of M components means simply an ordered array, as

$$\mathbf{V} = \begin{pmatrix} V_1 \\ \vdots \\ V_M \end{pmatrix},$$

whatever M is, and however, the V_j's are defined.

A vector may be defined as comprised of the hypothetical values of distinct variables whose observation is anticipated, such as

$$\begin{pmatrix} \text{pressure} \\ \text{temperature} \\ \text{salinity} \\ CO_2 \text{ conc.} \end{pmatrix}.$$

It may be defined as an ordered array of recorded values for a single, scalar variable at distinct times or locations, such as

$$\begin{pmatrix} \vartheta_{100} \\ \vdots \\ \vartheta_{900} \end{pmatrix},$$

where ϑ is the virtual potential temperature, for which we may have recorded values at standard pressure levels (denoted here in mb). Or it may be an ordered array of coefficients in a regression analysis. In all cases the interpretation of $\mathbf{V}$ is simply that of a matrix array where the matrix has one column and, say, M rows.

Occasionally in the text we will use the notation **VIP** to draw attention to a very important point.

EXERCISES

1. Write a statement of your objective in completing this course or reading this book. Connect it with what you wish to do, or are doing in your scientific career. (Get beyond the immediate goal of satisfying a course requirement!)

2. Focus on a scientific question which is of interest to you personally and which you conceivably might pursue.
   (a) State this question as it first came to mind, or as you might put it in conversation with your mom or a nonscientific friend.
   (b) Identify hypothetical or real data which you would use in arriving at an answer to this question.
   (c) Describe uncertainty and measurement errors connected with the collection and recording of the data, as these may affect the certainty of an answer to the question.
   (d) Say how you would expect to use the data to reach a conclusion.
   (e) Now restate the initial question, taking into account (b)–(d).

3. Refer to Question 3 in Table 1.1. Consider the data you would need and could realistically expect to obtain, to arrive at an answer. Keep in mind that most tornadoes over land are observed and reported, while those over oceans may not be, and that maximum wind velocity, which is a common indicator of the intensity of a hurricane, varies throughout the storm's lifetime.
   (a) In view of practical constraints restate the question as an answerable scientific question.
   (b) Identify the data sets you will need to create or access, in order to answer this question.
   (c) Describe uncertainty and measurement errors which may affect interpretation of the data relative to the (restated) question.

4. Refer to Question 4 in Table 1.1. With the objective of stating a research hypothesis concerning the effect of urban air pollution on urban climate, propose measurable variables which you might use as quantitative indicators of cause and effect.
   (a) Identify these variables; and describe both a sampling array and a sampling schedule for obtaining observed values to evaluate your hypothesis.
   (b) Think about and describe how the duration of your sampling program is related to variability of the urban atmosphere as well as to equipment and personnel errors.
   (c) Now state a research hypothesis in terms of the variables you have identified.
   (d) State the converse of the research hypothesis, also in terms of these variables. This should be a statement negating the effect hypothesized in part (c).

5. Refer to Question 5 in Table 1.1. Provide advice to the author of this question on a point, or points, that require clarification before a scientific program can be defined to answer the question.
   (a) Make your advice specific, in the form of a restatement of this question.
   (b) Outline the contents of a report which would contain information satisfying the proposer of the question. Include the identification of variables for which you wish to have measurements and a description of the extent and detail of (possibly hypothetical) data records.

6. Refer to Question 6 in Table 1.1 and answer the following.
   (a) How would you make this question sufficiently specific to answer with data from a research cruise? Rewrite it to satisfy this requirement.
   (b) Describe an array of bathythermographs which might be deployed to collect this data.
   (c) What circumstances might create ambiguity in the construction of an answer to the (revised) research question? How would these show up in the data?
   (d) What would you do to "have a first look at your data", to determine whether the bathythermographs were all functioning properly and whether the vertical temperature and salinity profiles may be considered to be homogeneous within the extent of your sampling array?

7. Refer to Question 7 in Table 1.1.
   (a) Turn this question into a pair of competing hypotheses. If you believe this is necessary, make them more specific. Explain your reasoning.
   (b) Describe the data you would use to answer this question, giving specific times, dates, and location(s).
   (c) Identify the statistic which you think will provide the most powerful discriminator for the two hypotheses; pick a (arbitrary) possible value for this statistic and state the conclusion you would reach if the statistic calculated from the data had this value. Give a brief argument in support of your conclusion.

8. Advise the author of Question 8 in Table 1.1 concerning this statement of a research question.
   (a) What does he need to be more specific about if he is to pose an "answerable question" in terms of relevant variables that can be measured. Hint: What variables are relevant?
   (b) Describe a hypothetical field experiment which you could conduct to arrive at an answer to this question. Include specification of the measurements/readings you would obtain in this investigation.
   (c) In the context of your answers to parts (a) and (b), describe how you might present collected evidence as an answer.

2

DATA AND
DATA MANAGEMENT
"what we have to go on"
or "accumulated records
of observations and their
expeditious reorganization"

2.1 INTRODUCTION

In this chapter we describe the data sets on the disk which accompanies this book. Examples that illustrate the text have been drawn from these data sets and many of the homework problems will use them. **Before inserting the disk into your PC's $3\frac{1}{2}''$ high-density disk drive, be certain that the "read only" window on the disk is open**. This will protect against the accidental loss of any of the data. When you do insert your disk into your PC, you may be asked whether you wish to initialize it. The answer is **NO**. Initialization or any other writing exercise will cause loss of data. The disk is filled to capacity; and all of the data will be referenced during your course of study.

There are four data sets in ASCII format on this disk. The data sets can be copied from the disk into your PC or local computer, either one at a time or all at once for permanent file storage. If you expect to read repeatedly from the disk, it is recommended that you make a copy of it now, for safekeeping. ASCII format is universally readable. However, your facility will undoubtedly require character translation and the creation of new files to be used for analysis. The following instructions assume that

MINITAB is available. Thus new files should be created in the format required by MINITAB on your system. Help with this step should be provided by your instructor or a computing facility consultant.

2.2 DAILY CLIMATOLOGICAL DATA FOR 15 STATIONS

This first data set, called CLIMAT on your disk, contains daily observations of several variables for 380 consecutive days beginning September 21, 1989. The stations from which the data were obtained are those climatological observing stations of the North American network identified in Fig. 2.1. Their selection gives interesting geographic diversity to this surface observation data set, which will be a focal point of some of your analysis exercises.

The first 15 lines of CLIMAT provide station locations and data inventories. One line for each station contains a three-letter station identifier, the station's coordinates (latitude, longitude, elevation), and the numbers of valid reports for each of the variables in the data set. The location and inventory file is reproduced here as the body of Table 2.1, i.e., as the portion of the table below all the column headings. Note the

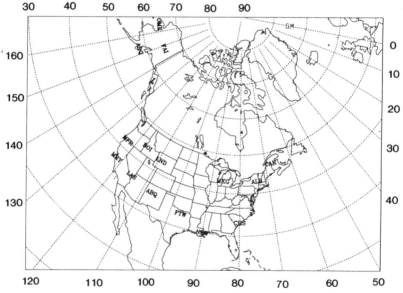

FIGURE 2.1   Locations of the climatological observing stations for which the data set CLIMAT contains observation records.

TABLE 2.1 Station ID, Physical Coordinates, and Inventories for the 15 Stations of the Daily Climatological Data Set CLIMAT

	Coordinates			Numbers of observations								
STN ID	LAT (°N)	LON (°E)	ELEV (m)	TOT PCP	AVG PRS	AVG T	MAX T	MIN T	AVG TD	AVG WDR	AVG WSP	PK WSP
C: 1–4	5–11	12–19	20–26	27–30	31–34	35–38	39–42	43–46	47–50	51–54	55–58	59–62
MRY	36.58	−121.85	67	0	0	332	332	332	332	334	334	334
LAS	36.08	−115.17	664	380	380	380	380	380	380	380	380	380
ABQ	35.05	−106.62	1620	380	380	380	380	380	380	380	380	380
FTW	32.82	−97.35	211	380	380	380	380	380	380	380	380	380
NEW	30.03	−90.03	3	380	380	380	380	380	380	380	380	380
CHS	32.90	−80.03	15	380	380	380	380	380	380	380	380	380
MFR	42.37	−122.87	405	380	380	380	380	380	380	380	380	380
BOI	43.57	−116.22	871	380	380	380	380	380	380	380	380	380
LND	42.82	−108.73	1594	380	380	380	380	380	380	380	380	380
MKG	43.17	−86.27	193	380	380	380	380	380	380	380	380	380
ALB	42.75	−73.80	89	380	380	380	380	380	380	380	380	380
CAR	46.87	−68.02	16	380	380	380	380	380	380	380	380	380
OME	64.50	−165.45	7	373	373	373	373	373	373	373	373	373
FAI	64.82	−147.87	138	373	373	373	373	373	373	373	373	373
ADQ	57.75	−152.50	34	373	373	373	373	373	373	373	373	373

TABLE 2.2   Variables and Format for Data of the 15 Stations of CLIMAT

Variable	Units	Format
Date: year, month, day	—	14, 2I3
Station identifier	—	A4
Total precipitation,	in.	F7.2
as liquid H_2O or equivalent		
Average atmospheric pressure	mb	F7.2
Average air temperature	°F	F7.2
Maximum air temperature	°F	F7.2
Minimum air temperature	°F	F7.2
Average dewpoint temperature	°F	F7.2
Average wind direction	Degrees, clockwise	F7.2
	from north	
Average wind speed	Knots	F7.2
Peak wind speed	Knots	F7.2

All data are right justified within allocated fields; and missing values are coded as "9999.00".

column designations above the body of the table for column format guidance.

Following the 15 lines giving locations and inventories are 380 (long) records of daily data. Each record begins with the date given as year, month, day. For example, "1989   9 21" is the date designation of September 21, 1989, in FORTRAN format I4, 2I3. After the date, are the station identifiers and their observation reports for that day, in FORTRAN format 15(A4, 9F7.2). Variables of the data set, their units, and the format with which they were written are shown in Table 2.2.

2.3 OCEAN TEMPERATURE AND SALINITY DATA

This short data set, called STN#27 on your disk, contains two subfiles which report average temperatures and salinities, for an oceanographic station on the Grand Banks of Newfoundland (see Fig. 2.2). The first subfile begins with a line which reads: "Temperature by month and depth for Station 27". This is followed by 10 lines containing the data reproduced in the body of Table 2.3a. These values were computed from an archive with reports for the years 1946–1991. For each depth and month, they are the arithmetic averages of all the observations in the archive for those years. The next line following the 10 lines of mean temperatures reads: "Salinity by month and depth for Station 27". This, in turn, is followed by 10 lines of data as shown in the body of Table 2.3b. Again these are arithmetic mean values calculated from the reports of Station 27,

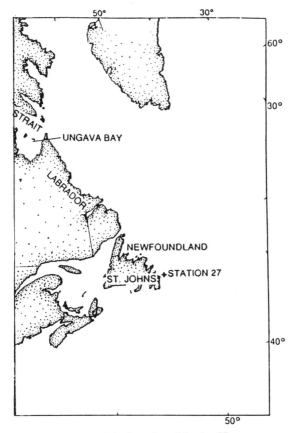

FIGURE 2.2 Location of Station 27.

for the years 1946–1991. Column designations and data formats for this file are as shown in the table.

Enterprising students who wish to have the full data set for more thorough exploration than that covered by this text may place a request with the author.

2.4 A 10-YEAR, DAILY TEMPERATURE, AND PRECIPITATION DATA SET AND 94 YEARS OF MONTHLY MEAN VALUES FOR STATE COLLEGE, PENNSYLVANIA

The next file on your disk, called STCOLL, contains data for the most recent 10 years of a 94-year data set which begins 1 January 1896. There is

TABLE 2.3a Station 27 Average Temperatures by Month and Depth, 1946–1991

Depth (m) C:	Jan 4–9	Feb 10–16	Mar 17–23	Apr 24–30	May 31–37	Jun 38–44	Jul 45–51	Aug 52–58	Sep 59–65	Oct 66–72	Nov 73–78	Dec 79–85
0	−0.19	−0.90	−1.04	−0.17	1.59	5.24	10.04	12.70	11.55	8.77	5.12	2.49
10	−0.24	−1.19	−1.23	−0.43	1.32	4.76	9.23	12.13	11.27	8.64	5.12	2.59
20	−0.26	−1.17	−1.26	−0.67	0.92	3.25	5.77	8.39	10.43	8.42	4.96	2.52
30	−0.23	−1.04	−1.24	−0.64	0.37	1.91	3.10	3.81	7.84	7.92	4.69	2.31
50	−0.30	−1.09	−1.32	−0.93	−0.55	−0.24	−0.25	0.10	2.04	4.90	3.83	1.98
75	−0.35	−1.08	−1.37	−1.23	−1.07	−1.07	−1.08	−0.96	−0.19	0.71	2.11	1.47
100	−0.33	−1.00	−1.36	−1.32	−1.32	−1.31	−1.36	−1.29	−1.03	−0.72	0.30	0.88
125	−0.51	−0.89	−1.26	−1.36	−1.37	−1.39	−1.42	−1.45	−1.28	−1.14	−0.85	−0.03
150	−0.42	−0.67	−1.02	−1.14	−1.18	−1.22	−1.29	−1.33	−1.23	−1.16	−1.05	−0.65
175	−0.53	−0.58	−0.78	−0.80	−0.85	−0.98	−1.08	−1.15	−1.07	−1.00	−0.99	−0.90

TABLE 2.3b Station 27 Average Salinities by Month and Depth, 1946–1991

Depth (m)	Jan	Feb	Mar	Apr	May	Jun	Jul	Aug	Sep	Oct	Nov	Dec
C:	4–9	10–16	17–23	24–30	31–37	38–44	45–51	52–58	59–65	66–72	73–78	79–85
0	32.03	32.23	32.36	32.24	32.10	31.96	31.62	31.26	31.07	31.09	31.34	31.74
10	32.03	32.27	32.36	32.35	32.16	31.99	31.73	31.32	30.98	31.07	31.35	31.71
20	32.07	32.30	32.38	32.39	32.25	32.10	31.97	31.65	31.12	31.11	31.40	31.73
30	32.09	32.29	32.40	32.42	32.33	32.30	32.24	32.08	31.42	31.21	31.48	31.81
50	32.17	32.38	32.44	32.52	32.55	32.62	32.62	32.57	32.17	31.65	31.73	31.93
75	32.32	32.45	32.55	32.52	32.71	32.78	32.83	32.79	32.61	32.39	32.17	32.16
100	32.50	32.58	32.64	32.75	32.83	32.91	32.94	32.91	32.83	32.73	32.58	32.42
125	32.70	32.74	32.80	32.85	32.95	32.98	33.00	32.99	32.98	32.91	32.87	32.69
150	32.97	32.93	32.91	32.99	33.04	33.09	33.10	33.09	33.10	33.09	33.07	32.95
175	33.18	33.07	33.06	33.10	33.15	33.20	33.22	33.21	33.25	33.27	33.29	33.17

TABLE 2.4   **Daily Temperature and Precipitation Data for State College, PA, January 1980**

	YEAR	MO	DAY	MAX T	MIN T	AVE T	PRCP	SNOW
C:	1–4	5–7	8–10	11–15	16–20	21–27	28–34	35–40
	1980	1	1	42	20	31.0	0.00	0.0
	1980	1	2	43	19	31.0	0.00	0.0
	1980	1	3	32	26	29.0	0.00	0.0
	1980	1	4	37	18	27.5	0.00	0.0
	1980	1	5	23	19	21.0	0.16	1.4
	1980	1	6	27	13	20.0	0.00	0.0
	1980	1	7	33	11	22.0	0.01	0.0
	1980	1	8	41	19	30.0	0.01	0.1
	1980	1	9	31	19	25.0	0.00	0.0
	1980	1	10	29	13	21.0	0.06	0.8
	1980	1	11	34	12	23.0	0.00	0.0
	1980	1	12	47	22	34.5	0.12	0.0
	1980	1	13	27	14	20.5	0.00	0.0
	1980	1	14	32	13	22.5	0.13	0.0
	1980	1	15	38	31	34.5	0.64	0.0
	1980	1	16	45	26	35.5	0.00	0.0
	1980	1	17	45	23	34.0	0.00	0.0
	1980	1	18	40	36	38.0	0.00	0.0
	1980	1	19	42	34	38.0	0.00	0.0
	1980	1	20	34	31	32.5	0.00	0.0
	1980	1	21	37	24	30.5	0.00	0.0
	1980	1	22	35	25	30.0	0.00	0.0
	1980	1	23	38	31	34.5	0.02	0.1
	1980	1	24	32	9	20.5	0.00	0.0
	1980	1	25	18	9	13.5	0.06	1.5
	1980	1	26	34	17	25.5	0.02	0.3
	1980	1	27	30	13	21.5	0.00	0.0
	1980	1	28	33	13	23.0	0.00	0.0
	1980	1	29	30	18	24.0	0.00	0.0
	1980	1	30	25	10	17.5	0.00	0.0
	1980	1	31	25	11	18.0	0.01	0.1

one record for each day of the 10 years 1980–1989,[1] in the file STCOLL. Table 2.4 reproduces the data for January 1980, and shows the column format of the data as it is stored on your disk. Special note should be made of entries of " − 1.00" in the precipitation or snow columns. These signify what is commonly called "a trace". The value of 0.00 means "none"; measurable amounts are reported to two decimals in the case of

---

[1]Each of the 10 years has a 29Feb row. For the 7 years which are not leap years, a string of four 9's fills each data entry space: "9999" for the integer-valued variables and "9999" for variables with reported decimals.

liquid water or its equivalent, and snow depths are reported to one decimal place.

The second file for State College is called 94YEARS. This one contains monthly means of daily maximum, minimum, and average temperature values. The file has one record for each year, with the year number as the lead element in each record. This is followed by the data in sets of 13 numbers: JAN, FEB, ..., DEC, and YEAR MEAN. The first set contains the mean values for daily maximum temperatures, the second for daily minimum temperatures, and the third for daily average temperatures. The records were written with FORTRAN format I4, 39F6.2.

2.5 DAILY MIN / MAX TEMPERATURE TIME SERIES FOR FIVE CITIES

The data set called TEMPS on your disk contains the most recent 37 years' data of an 88-year archive of daily minimum and maximum temperatures for the five cities:

Crookston, Minnesota (CRKS)

Little Rock, Arkansas (LIT)

New York City (NYC)

Saginaw, Michigan (SAG)

State College, Pennsylvania (STC)

The dates for the records on your disk are 1 January 1950 through 31 December 1986. Each record in the file gives the date and five successive pairs of values: one MIN and one MAX temperature report for each city in the order listed above. Table 2.5 reproduces the data for the first month, with the column and data formats used throughout the file.

"Daily average temperatures", which you will be using in the exercises, are computed as the arithmetic averages of the daily MIN and MAX values. You may wish to make this simple calculation and add the five columns of daily average temperatures to your data set, at the time you translate and store the MIN and MAX temperature values at your computing facility. One of the exercises in the next section will guide you in this construction.

2.6 USING MINITAB FOR DATA MANAGEMENT

In this section we present a brief description of data management using MINITAB, to get you started and comfortable with the ease of data selection and the definition of new arrays. To go beyond the simple tasks

TABLE 2.5 Daily Minimum and Maximum Temperatures for the Five Cities

YEAR	MO	DY	CRKSMIN	CRKSMAX	LITMIN	LITMAX	NYCMIN	NYCMAX	SAGMIN	SAGMAX	STCMIN	STCMAX
1-4	5-7	8-10	11-15	16-20	21-25	26-30	31-35	36-40	41-45	46-50	51-55	56-60
1950	1	1	-6	25	54	63	33	42	35	41	33	40
1950	1	2	-9	0	60	70	39	43	41	46	35	42
1950	1	3	-19	-3	67	72	43	60	46	55	38	55
1950	1	4	-24	-15	23	71	59	66	20	52	54	65
1950	1	5	-22	1	24	31	48	64	18	24	43	56
1950	1	6	-22	0	30	37	45	63	20	26	38	51
1950	1	7	-32	-1	28	45	31	45	11	29	26	40
1950	1	8	-7	9	28	45	16	31	8	30	16	30
1950	1	9	-10	11	40	65	24	45	29	44	22	51
1950	1	10	-17	5	44	65	42	54	23	52	40	60
1950	1	11	-20	25	38	47	29	55	12	25	22	42
1950	1	12	-5	25	43	68	27	36	18	40	21	29
1950	1	13	-16	0	53	69	36	46	34	54	29	38
1950	1	14	-22	-7	43	56	40	61	19	49	32	55
1950	1	15	-15	-4	44	75	33	51	19	42	23	38
1950	1	16	-26	-13	34	50	36	52	11	21	28	44
1950	1	17	-24	-10	39	50	30	39	10	37	20	35
1950	1	18	-34	-12	30	74	35	49	3	37	23	49
1950	1	19	-29	-10	29	36	24	33	3	20	17	25
1950	1	20	-18	5	28	44	19	32	5	21	15	27
1950	1	21	-10	10	38	55	22	37	18	32	22	35
1950	1	22	-4	10	52	66	34	50	30	35	30	40
1950	1	23	-9	1	56	75	44	54	22	30	40	52
1950	1	24	-11	17	68	81	41	49	25	51	41	55
1950	1	25	-28	-9	68	83	36	41	38	62	45	71
1950	1	26	-31	-17	32	75	39	72	11	55	37	66
1950	1	27	-27	-7	28	45	28	58	9	23	23	37
1950	1	28	-17	0	39	60	24	40	20	39	19	35
1950	1	29	-31	-5	36	75	40	53	13	39	35	54
1950	1	30	-33	-11	30	36	34	51	5	17	25	51
1950	1	31	-21	-8	34	36	30	35	12	28	24	35

described here and to do the numerical exercises at the end of the chapters, we recommended that you obtain a copy of the *MINITAB Handbook* (Ryan et al., 1992) from your campus bookstore. Although the examples in the handbook are not drawn from physical science, it will be an excellent companion to your text on statistical analysis and valuable as a future reference. It is well worth the investment of its modest cost to have it at hand.

The description we present will take the form of worked examples using one of the data sets on your disk. First we will go through the steps of creating a subset of the data according to a numerically quantifiable criterion, naming it for later reference and saving it as a MINITAB file. Then we will create a totally new file, with a simple arithmetic operation on columns of the original MINITAB file, followed by intrafile transfers and deletion of unnecessary data; and we will save this as well. Clearly these maneuvers assume that the data set is already in MINITAB-accessible format.

Creating MINITAB files from the ASCII files on your disk will require the assistance of someone familiar with your local computing facility. It can be done with a simple read/write command; however, the language of the command will be specific to the computer you will be using. A computing consultant can explain the allocation of file space and the read/write command which transfers the data into MINITAB file format, at the same time that she or he shows you how to log on and off the machine, and how to access the MINITAB software.

Our example uses the data set called TEMPS. Step 1 is to create and name the MINITAB file with this data in it. Let's assume that you have accomplished this and that you have named it TEMPS also. Although it may seem a bit confusing at first, in the process of reformating files you will have discovered that in MINITAB jargon, a column designation is an array name. This is in distinction to the single-place columns of ASCII and FORTRAN format statements. Thus, in the language of MINITAB, a "column" is an array containing the successive values of one of the variables of the data set. For example, the original MINITAB file TEMPS will have 13 columns: C1 containing the values of YR, C2 containing the values of MO,..., C12 containing the values of STCMAX.

If you can sit at a terminal or PC to apply the following steps as you read about them, the data maneuvers will be much easier to learn. Let's assume that you can and that you have accessed an interactive version of MINITAB, however that is done on your computer. Once there, you will discover that the rest is amazingly easy.

The first thing we can do is "retrieve" TEMPS from the space allocated for the retention of files by the computer, with a command that will put a copy of this file into what we call MINITAB's worksheet. You need not worry about damaging it irrevocably by what you do to the worksheet copy with MINITAB software. The original will not be written-over. If you mess up, you can retrieve another copy of it and start again: older and wiser and a little bit later than you wish it were.

To retrieve the file and confirm its retrieval into MINITAB's worksheet, follow these steps:

1. Get into the interactive mode with MINITAB. The software will respond with the prompt MTB⟩. In words, it awaits your command. To create a paper copy of everything you do with this exercise, for future reference, follow the prompt by typing OUTFILE 'EVENTS' and hit the return key. The software responds with another MTB⟩ prompt. This tells you that it did what you asked and now awaits your next command. An output file has been created, to which the software will add all the records of your current transactions with the computer, until you instruct it to cease accumulating information in this file with a NOOUT command.

2. Retrieve your data set by typing the command RETRIEVE, followed by the data set location/name in single quotes. The current line on your screen should now look like

MTB⟩ RETRIEVE 'location/name'

with the actual location/name specification for your data set substituted within the quotes. When you hit the return key, the software again responds with the prompt MTB⟩.

3. If you wish confirmation that the file has truly been retrieved and that a copy is available for your use, type the command PRINT followed by designations of the columns you wish to view. If you wish to see only the first five, your screen should look like

MTB⟩ PRINT C1-C5.

If you wish to see all the columns in the data set, your screen should look like

MTB⟩ PRINT C1-C13.

When you hit the return key, the software response will fill your screen with data from the columns you have specified and it will ask you if you want more. You respond by typing either Yes or No. Since there is too much data to run through it all, you should respond here with No Ⓡ. The software response will again be MTB⟩, indicating that it again awaits the next command.

Next, we can add descriptive information to our copy of the file by assigning to the columns the names of the variables whose values are stored in them, and then confirm our success. To do this we use the following:

4. Name the columns descriptively, using between 2 and 8 characters for each name. To use the names suggested by Table 2.5, you name the

first three by typing NAME C1 'YR', C2 'MO', C3 'DY' after the prompt. Alternatively you may wish to spell out these words fully, with a command line that looks like

MTB〉 NAME C1 'YEAR', C2 'MONTH', C3 'DAY'.

To name the next two columns, use the command line

MTB〉 NAME C4 'CRKSMIN', C5 'CRKSMAX'

etc., until finally you have

MTB〉 NAME C12 'STCMIN', C13 'STCMAX'.

5. To confirm that the columns have their new names, use

MTB〉 PRINT C1-C13.

When you hit the return key this time, the software response will fill your screen with the same data as before; however, in place of the column headings C1 C2...C13, will be the headings

YEAR MONTH...STCMAX.

Now, for everything you do with this file, you can refer to the columns by their new names if you wish to do so.

Creating and Saving a Subfile

The next thing we will do is to select the data for Crookston, Minnesota for the month of May 1986, and create a new file containing just the days of this month and the corresponding minimum and maximum temperatures for Crookston. We do this by first declaring that we are going to save a file with the name MAY.1986, which is accomplished with the command line

MTB〉 SAVE 'MAY.1986'

and then editing of the full file to create this much shorter one.

In its present form our copy of the file TEMPS has 13 columns. We can add columns to it, up to the total column limit of 999. (In other words, we are not likely to exceed the permissible number of columns!) If we wished to copy all of the temperature information for Crookston into new columns, say into columns 24 and 25, with the day of the month simultaneously copied into column 23, we would use the command line

MTB〉 COPY C3-C5 C23-C25.

You may wish to try this and confirm that it has been accomplished with

MTB〉 PRINT C3-C5, C23-C25.

Since we want the Crookston data only for month 5 of year 1986, we attach a subcommand. That is, we follow the COPY command line with a semicolon and hit the return key. The software responds with the prompt SUBC⟩, which tells you that it is ready for your qualifying instructions. In this example, the qualification to the COPY command is that we want to use only those lines of the file for which the year is 1986 and the month is 5. Accordingly, our copy and subcommand pair looks like

MTB⟩ COPY C3-C5 C23-C25;

SUBC⟩ USE 'YEAR' = 1986 AND 'MONTH' = 5.

Try this and then enter PRINT C23-C25 ⓇR to confirm that it has worked.

VIP: It is essential to get the punctuation right for this to work. If your entries have not produced the desired result, carefully review what is on your screen. There must be a space between the designations of the columns you are copying from and the columns you are copying into. The semicolon must follow the copy command to signal the software that it will receive a qualifying subcommand. If you use names rather than column numbers in the subcommand, the names must be in single quotes. And, finally, the subcommand must end with a period. (With a one-line command, the period is unnecessary. However, here where there is a qualification to the command, it must be used to signal the software that it has reached the end of the instructions.)

When you have confirmed that you have precisely the values you want in columns 23 through 25, you can copy these columns onto the first three columns of your worksheet, with the command line

MTB⟩ COPY C23-C25 C1-C3.

Now erase the columns of the worksheet that you no longer require, with

MTB⟩ ERASE C4-C999.

One final edit maneuver will give your new, smaller file the appearance you wish it to have: rename the columns with the names of their present contents. MINITAB does not tolerate duplicate names and 'DAY' is the current name for column 3, so we must perform the renaming out of sequence. First use

MTB⟩ NAME C2 'CRKSMIN', C3 'CRKSMAX'

and then

MTB⟩ NAME C1 'DAY'.

Now if you give the command to print columns 1 through 3, the response on your screen should be DAY CRKSMIN CRKSMAX, followed by 31 lines, each with the day of the month and the day's minimum and maximum temperatures for the 31 days of May 1986. If not, go back and critically review your command lines. Without a doubt you will find some small detail of punctuation, which you can correct to make your next run through this exercise successful.

What you have done is to create a new data set, called MAY.1986; and you have already told MINITAB that you want it saved. You have also created several screens of command lines and responses. It is highly likely that these include some retrys of commands that were not put in correctly the first time, and they comprise a valuable record for study in preparation for your next MINITAB encounter. If you gave the OUTFILE 'EVENTS' command at the start, that record has been created for you. All you need to do now is to give the NOOUT command, get out of MINITAB and back onto your local system, and print both files: MAY.1986 and EVENTS. So, use

MTB⟩ NOOUT

MTB⟩ STOP

and ask your computer consultant how to print the two files which are now stored in your assigned computer space.

Creating a New File from the Contents of an Original

MINITAB can do some simple arithmetic operations with direct commands, in the creation of new columns in the worksheet copy. We will exercise this option by computing average temperatures for the five locations whose minimum and maximum temperature records comprise columns 4 through 13 of the TEMPS file; and we will save the result as AVETEMP. Again retrieve a copy of TEMPS into the MINITAB worksheet. You can specify that you want to create and save a new file called AVETEMP at the outset, with the command line

MTB⟩ SAVE 'AVETEMP'.

Since the daily average temperatures will be the simple arithmetic averages of the daily minimum and maximum temperatures, we add the pairs of numbers and divide by two, and store the results in new columns. MINITAB's LET command permits us to do this. Again, only the first thirteen columns have data in them, so we may put our results in any later columns we choose. Anticipating copying them over, it will be simpler if we choose consecutive columns, although this is not required by the software. Recall that the date uses three columns, C1-C3, which we wish to leave undisturbed in this case. Beyond these three columns, we wish to average consecutive pairs of values. This is accomplished with the command lines

MTB⟩ LET C21 = (C4 + C5)/2
MTB⟩ LET C22 = (C6 + C7)/2
MTB⟩ LET C23 = (C8 + C9)/2
MTB⟩ LET C24 = (C10 + C11)/2
MTB⟩ LET C25 = (C12 + C13)/2

which store the results in columns 21 through 25. That is, we have the full sequence of daily average temperatures for Crookston in column 21, for Little Rock in column 22, for New York in column 23, for Saginaw in column 24, and for State College in column 25. We can now copy these over the old entries in columns 4 through 8, with

MTB⟩ COPY C21-C25 C4-C8

and scrap the remaining columns, with

MTB⟩ ERASE C9-C999.

Now the file AVETEMP contains only the dates and the corresponding daily average temperatures for our five locations. You can name these columns if you wish, with the naming instructions; and you can confirm that the values look like what you anticipate with a PRINT C1-C8 command. When you exit MINITAB with

MTB⟩ STOP

your new file will be saved in the space allocated to you by the computer.

A Word of Caution

You should not try to print this new file in its entirety. Since the original TEMPS file had 37 years of data, your new one will as well, if you have made it correctly. If you wish to see part of it, you can either use an OUTFILE command and PRINT part of it to the screen, followed by NOOUT; or you can select and save some part of it using MINITAB commands, much as we did in creating and saving MAY.1986, and you can print that smaller file with the system printer.

EXERCISES

1. Get into MINITAB and retrieve AVETEMP, the file that was created and saved in the second demonstration exercise. Make a subfile of the daily average temperatures of Crookston, for only the days of May 1986. Obtain a printed copy of this by one of the two options. Compare it with your printed copy of the daily minimum and maximum temperatures of the file called MAY.1986, for confirmation that you have what you think you should have.

2. Start with the original data set TEMPS and create a new file which contains the day of the month in column 1, and the following information for State College, in columns 2–7.

   (a) For January 1980, daily minimum, maximum, and average temperatures.

   (b) For July 1980, daily minimum, maximum, and average temperatures.

   Name all columns descriptively. Save the new file and print it.

3. (a) Start with the original data set TEMPS and create a new file which contains the dates and the daily temperature ranges: maximum-minus–minimum, for each of the five locations.

   (b) Into columns in sets of five, further out in the file, copy the daily temperature range values for (i) January 1950, (ii) January 1960, (iii) January 1970, and (iv) January 1980, and name these columns descriptively.

   (c) SET the values $1, 2, \ldots, 31$ into C1 and name this column 'JAN-DAY'. You can either do this directly with the command lines

   MTB⟩ SET C1
         1 2 3 4 5 6 7 8 9 10 11 12 13 14 15 16 17
         18 19 20 21 22 23 24 25 26 27 28 29 30 31
         END

   or you can COPY the original C3 into C14, say, using only C1 = 1950 and C2 = 1. If you do the latter, then you will need to copy C14 onto the original C1 and ERASE C14.

   (d) ERASE the columns of the file containing the complete lists of temperature ranges, i.e., those for all 37 years, and the 'MONTH' and 'DAY' columns, i.e., the original C2 and C3. Now copy the decadal January temperature ranges into columns 2–6, 7–11, 12–16, 17–21, and print the result. To make it easier to read, separate the sections of your table that pertain to the four different years with vertical lines drawn by hand, and print section headings and a single table heading overall.

4. (a) Create two MINITAB files from the data in STN#27, one with monthly average temperatures and one with monthly average salinities. Name the columns and print these files.

   (b) Starting with copies of the MINITAB versions of the temperature and salinity files, create two new files. These files will have in their columns the changes in the average values since the *previous* month, starting with the Jan column which will contain Jan-minus-Dec values. Name the columns of these files descriptively and obtain printed copies.

(c) Make certain that the four tables you have created in parts (a) and (b) are labeled unequivocally.

5. Create a MINITAB file containing the data in CLIMAT. From a copy of the MINITAB file make a new file containing only the data for Boise (BOI). Add a column which contains daily temperature ranges, i.e., the daily values of maximum–minus–minimum temperatures. Name the columns; print the first 91 days of this file, so that you have a hard copy of the record for Fall of 1989; and save 'BOISE'.

6. Create a MINITAB file containing the data in STCOLL. From a copy of the MINITAB file make a file containing the dates, liquid precipitation values, and average temperatures, for Spring of 1985. Name the columns, save and print this file, and label the printed copy clearly.

3

DESCRIPTIVE STATISTICS
"first impressions" or "sketching features of observed systems with data"

3.1 INTRODUCTION

We use statistical tools for extraction and condensation of information relevant to our scientific objectives, from assembled data. Exploratory tools which will be discussed in this chapter offer a powerful beginning. Quite simply, they describe the data without reference to any hypotheses or assumptions we may bring to the project. Some of these statistics will be familar to you. All are given rigorous definition, with emphasis on the fact that their values are, purely, descriptors of the collection of data from which they have been calculated. Preliminary examination of a data set with these statistics can be an invaluable aid to further detailed analysis. Whether the format recommended in Chapter 1 has been followed in the planning stage, or data have been assembled prior to giving detailed thought to their analysis, this "first look" focuses on the analysis starting point. It may reveal features of the data which were not anticipated and, in doing so, suggest relationships which must be taken into account in completing the analysis. We save significant time and energy in knowing about these characteristics of the data at the outset.

We will be using computer software which makes it extremely easy to do preliminary examinations of data files. Perhaps the best known software packages which offer easy construction of statistical profiles are MINITAB, SAS, and S-PLUS. The illustrations in this and later chapters provide examples of output from some of their routines. Generally when there is a large volume of data in the file, SAS or S-PLUS will be easier to use than MINITAB in carrying out a complete analysis. For relatively short, uncomplicated data files, MINITAB is wondrously easy to understand and use. Thus it is a valuable learning tool. In all cases and at all computation centers, available statistical analysis software makes data exploration extremely easy in contrast with older, hand calculation meth-

TABLE 3.1 Daily Surface Station Data for State College, PA, October 1990

Date	Pressure (in. Hg)	Avg. temp. (°F)	Max − min temp. (°F)	Dewpoint temp. (°F)	Relative humidity (%)	Avg. wind spd. (mph)	Peak wind spd. (mph)
1	30.0	55.4	19	39.3	56	5.6	25
2	30.2	51.0	24	35.2	57	4.1	19
3	30.2	60.1	26	44.4	57	5.6	17
4	29.9	56.9	13	49.7	77	6.2	19
5	30.1	63.7	17	44.1	50	5.2	19
6	30.1	66.7	28	46.0	48	4.1	16
7	30.2	67.0	25	50.5	55	2.2	12
8	30.1	68.1	17	57.5	68	1.0	10
9	30.4	70.2	16	60.5	71	3.2	13
10	—	—	—	—	—	—	—
11	30.1	56.6	15	53.0	88	1.6	18
12	30.0	65.1	17	59.4	81	1.2	12
13	29.8	65.1	15	57.8	77	1.6	8
14	29.9	58.2	62	46.6	66	2.5	13
15	30.1	52.3	22	35.4	55	4.7	17
16	30.3	53.9	20	36.2	53	1.8	9
17	30.1	63.7	8	48.1	57	5.6	18
18	29.7	45.6	26	34.5	67	12.2	31
19	30.2	38.1	19	27.8	67	4.5	21
20	30.3	47.7	26	31.2	56	2.4	10
21	30.2	53.2	17	42.2	68	2.9	12
22	30.0	54.3	9	50.9	89	0.2	6
23	29.8	54.6	10	50.1	85	1.9	14
24	29.9	48.1	55	36.9	67	2.2	11
25	29.9	46.6	20	34.0	64	3.9	16
26	30.1	34.6	17	20.1	60	3.3	17
27	30.1	44.7	25	26.0	51	3.5	18
28	30.1	38.9	13	26.0	63	9.0	21
29	30.3	38.4	21	22.4	56	2.5	13
30	30.2	54.8	32	28.6	39	3.6	12
31	30.2	53.6	20	37.0	55	0.2	5

ods. The output is highly presentable and provides unambiguous descriptions of the data to which the computational packages have been applied. Since MINITAB is the easiest to learn and use as an aid to course work, we recommend that you acquaint yourself with its commands and apply them to a familar data set. This will give you confidence as you approach the exercises and, ultimately, with data analysis in your own work.

There is always a root objective in measuring and recording the values in a data set. However, scientific interest is generally not in the individual values, but in the collective description they provide of the system from which they are obtained. Together they characterize that system: its range of values, variability, coherence, etc. Examination of a data set as a collection of numbers in a table, such as the surface weather data in Table 3.1, will yield general ideas about collective characteristics of the variables represented, provided the data set is available in hard copy and is sufficiently small to examine in this manner. However, it is difficult to get a well-focused impression from such a collection of numbers. Imagine that you have been asked to review the information in the table and to report on its noteable features. With pencil in hand, you might underline extreme values, bracket successions of like values, highlight central values, note the difference between the largest and smallest values for each variable, etc. In other words, what we seek initially are summary statistics that express these characteristics of the data records for each variable. This information, together with plots of the time series of recorded values,

TABLE 3.2 Descriptive Statistics for Surface Station Data for State College, PA, October 1990

	N	MEAN	MEDIAN	TRMEAN	STDEV	SEMEAN
PRESSURE	30	30.083	30.100	30.088	0.166	0.030
AVETEMP	30	54.24	54.45	54.47	9.64	1.76
DELT	30	21.80	19.50	20.00	11.53	2.10
DPT	30	41.05	40.75	41.12	11.39	2.08
RELHUM	30	63.43	61.50	63.04	12.23	2.23
WINDSPD	30	3.617	3.250	3.342	2.528	0.462
PEAKW	30	15.07	15.00	14.81	5.56	1.02

	MIN	MAX	Q1	Q3
PRESSURE	29.700	30.400	29.975	30.200
AVETEMP	34.60	70.20	47.42	63.70
DELT	8.00	62.00	15.75	25.25
DPT	20.10	60.50	33.30	50.20
RELHUM	39.00	89.00	55.00	68.75
WINDSPD	0.200	12.200	1.875	4.825
PEAKW	5.00	31.00	11.75	18.25

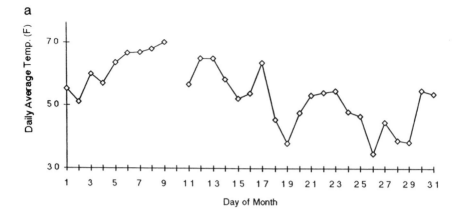

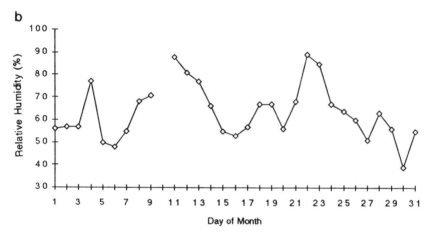

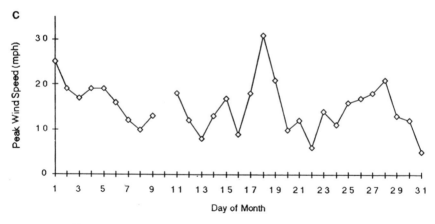

FIGURE 3.1 (a) Daily average temperatures, (b) relative humidities, and (c) peak wind speeds, for State College, PA, October 1990.

gives a detailed profile of the data set. Fortunately for us, working in the computer age, most data are maintained in computer accessible files and sophisticated software exists to give us summary information and visual displays quickly and easily. This support technology is universally available and makes a breeze of statistical computations and data plots that were formerly arduous or prohibitively time consuming. For example, with the data in Table 3.1 in a MINITAB file, the command DESCRIBE C2-C8 produced output shown in Table 3.2, and the command PLOT C3 VS C1, C6 VS C1, C8 VS C1 created the results reproduced in Fig. 3.1. With very little effort and the aid of the ubiquitous computer, we have much more detailed descriptions of data records for these variables than was evident from their tabular presentation.

3.2 DATA PLOTS, HISTOGRAMS, AND FREQUENCY DISTRIBUTIONS FOR A TIME SEQUENCE OF OBSERVATIONS

Many properties of time records of observations are easily discerned by viewing plots of their values, i.e., with time or date on the abscissa. Figures 3.1(a)–3.1(c), which illustrate records of daily average temperatures, relative humidities, and peak wind speeds, provide examples. A time sequence plot creates a mental image of the record which has information in a form which is generally more useable than an image produced by a column of numerical values. Without the precision of all the recorded digits, it conveys the range, variability, sequential relationships, and any evidence of periodicities in the data, as well as highlighting any values which are outliers and thus may be suspected of being in error. A time sequence plot of the original data contains much more information than a summary plot or descriptive statistics, because *it is the record* in distinction to any condensation of it. Accordingly, unless a sequential data set is very long, it is highly recommended that its time sequence plot be examined before further analysis is undertaken. Usually examination of this plot will reveal characteristics of the data set on which you will wish to focus attention via further analyses, as well as enabling you to identify and remove or repair any erroneous values that would lead to spurious conclusions if they contributed to statistical summaries.

Summary plots and descriptive statistics serve to reduce the detailed information in the original record to a few pieces of information that describe characteristics of that record, when full detail is neither required nor desired. In this section, some of the most useful summary plots are introduced using standard notation, with notes on the rationale for their prominence in statistical data analysis.

The most frequently encountered summary plot is a *histogram* of the values in the data set. If we were to construct this longhand, we would pick a number of consecutive intervals of equal length which span the range of values in the set, and tabulate the numbers of data values which fall within each interval. The number of intervals selected involves a trade-off between precision and useful summarization. Usually between 5 and 15 intervals will serve your purpose well, although some experimentation with your data is advisable to achieve a satisfying display. Having tabulated the frequencies of observations for each of the consecutive intervals, a histogram is constructed by drawing a bar graph with frequency as the ordinate and the midpoint of each interval marked on the abscissa. MINITAB can do this for you, with the values in any column of a MINITAB data set. For example, with the data of Table 3.1, HISTOGRAM C3, C6, C8 produced the results shown in Fig. 3.2, using its default interval selector.

The appearance of the distribution of observed values is highly dependent on interval choice; and you may find that MINITAB's automatic selection does not serve your display requirements. The software gives you the option of making the selection yourself by specifying INCREMENT which is the length of each successive interval and START which is the midpoint of the leftmost interval. It may be advisable to experiment with these, until you are satisfied with the visual resolution of the data display. Fortunately, with computer technology, this is extremely easy. To see the impact of interval choice, consider Figs. 3.3(a)–3.3(c), which are reformulations of Fig. 3.2(b), produced with the commands given in the figure caption. Although all three truly represent the same data, we note that if we choose intervals that are either very short, as in Fig. 3.3(a), or very long, as in 3.3(c), the resulting displays are unsatisfying. Comparison of Figs. 3.2(b) and 3.3(b), which both provide good balance between precision and summarization, brings home the point that there is no one "correct" histogram format for a data set. The choice is a matter of personal intuition and esthetics.

Common alternatives to presentation of the information in a frequency histogram are constructions of a "relative frequency distribution" or a "cumulative frequency distribution", which also provide visual displays of the frequencies of recorded values. The *relative frequency distribution* (r.f.d.) is constructed in the same way as the frequency histogram just described but with each interval count divided by the total number of data points before plotting. Thus the ordinate gives the fraction of the data set within each interval or "relative frequency of representation of that interval". The sum of these frequencies is always 1. When data sets of different sizes are to be compared with one another, or a hypothetical distribution is to be compared with the distribution exhibited by a data set, r.f.d. plots provide the appropriate bases, because the relative frequencies are comparable.

a
Histogram of AVETEMP   N = 30

```
Midpoint    Count
   35         1    *
   40         3    ***
   45         3    ***
   50         4    ****
   55         9    *********
   60         2    **
   65         6    ******
   70         2    **
```

b
Histogram of RELHUM   N = 30

```
Midpoint    Count
   40         1    *
   45         0
   50         3    ***
   55        10    **********
   60         1    *
   65         6    ******
   70         3    ***
   75         2    **
   80         1    *
   85         1    *
   90         2    **
```

c
Histogram of PEAKW   N = 30

```
Midpoint    Count
    4         1    *
    8         3    ***
   12        10    **********
   16         6    ******
   20         8    ********
   24         1    *
   28         0
   32         1    *
```

FIGURE 3.2 Frequency histograms for October 1990 for State College: (a) Daily average temperatures, (b) relative humidities, and (c) peak wind speeds.

```
35.00  0
37.20  0
39.40  1  *
41.60  0
43.80  0
46.00  0
48.20  1  *
50.40  2  **
52.60  1  *
54.80  3  ***
57.00  6  ******
59.20  1  *
61.40  0
63.60  2  **
65.80  1  *                                                                  .
68.00  5  *****
70.20  1  *
72.40  0
74.60  0                  40.00   1   *
76.80  2  **              47.50   3   ***
79.00  0                  55.00  10   **********
81.20  1  *              62.50   4   ****
83.40  0                  70.00   6   ******
85.60  1  *              77.50   3   ***          50.0  15  *****************
87.80  1  *              85.00   2   **           75.0  13  *************
90.00  1  *              92.50   1   *           100.0   2  **

a. mdpt.  count       b. mdpt.  count          c. mdpt.  count
```

FIGURE 3.3 Frequency histograms for October 1990 relative humidities for State College, made with the commands: (a) HISTOGRAM C3; INCREMENT = 2.2; START = 35. (b) HISTOGRAM C3; INCREMENT = 7.5; START = 40. (c) HISTOGRAM C3; INCREMENT = 25; START = 50.

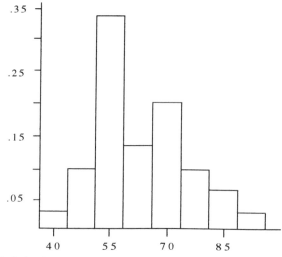

FIGURE 3.4 Relative frequency histogram corresponding to Fig. 3.3(b), with relative frequency on the vertical axis.

We construct a r.f.d. from a histogram simply by rescaling the vertical axis by the factor $1/N$, where N is the number of data points. For example, the plot in Fig. 3.4 is the relative frequency version of Fig. 3.3(b). Again we note the strong dependence of the visual image of the histogram on the choice of interval length and midpoints. When making comparisons between data sets with relative frequency histograms, it is necessary that these be identical. Comparable displays may be created using the abscissa of the first r.f.d. as a template.

The *cumulative frequency distribution* (c.f.d.) is a different type of display, generally serving a rather different purpose. With data values in the time sequence of their observation, the c.f.d. is created by first reordering the data to place the values in numerical order. For each value in the set, the cumulative frequency tabulation is the number of recorded values less than or equal to it. Clearly this will be 1 for the first value in the reordered set and N for the last of a set of N data, with an increase at each recorded value in between. An example is provided by Fig. 3.5(a), for the data of Figs. 3.3(b) and 3.4.

To establish the links between these frequency distributions and concepts of probabilities, as we shall do later, it is expedient to divide each tabulated value by the total number of observations. The resulting set of values and its graphical display readily translate to a monotonically increasing function,

$$F(x) = \text{proportion of recorded values} \leq x$$

as shown in Fig. 3.5(b). This has a value at every point on the axis of realizable values of the observed variable; and it increases from 0.0 on the left of the minimum value in the data set, to 1.0 at and to the right of the maximum value. In mathematical terminology, it is continuous from the right, with jump discontinuities at each distinct recorded value.

In our example there are repeated values in the data record. It is more likely that these reflect numerical truncation in recorded values than truly identical average temperatures for different days during the period of observation. Whatever their source, they show up in the c.f.d. as jumps of two and three units, by comparison with the unit jumps of unique values.

Note: Because the relative frequency distribution is based on intervals of possible values while the cumulative frequency distribution increases at each recorded value, the c.f.d., together with N, contains *all the information in the data set* except the order in with the values were obtained. Thus it provides a more complete description than the r.f.d. which has lost some of the precision in the original record. Nonetheless, both will be of extraordinary value in our exploration of techniques for statistical inference.

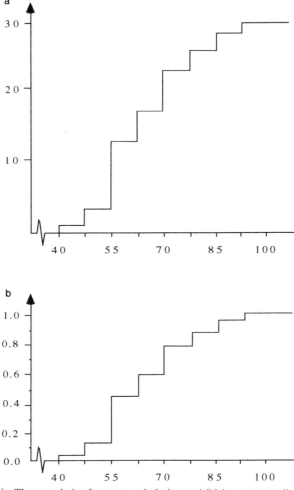

FIGURE 3.5 (a) The cumulative frequency tabulation and (b) its corresponding cumulative frequency distribution function.

3.3 SUMMARY STATISTICS

Plot, histogram, and frequency distribution functions of statistical analysis packages provide visual images of a data set not obtained from inspection of a collection of numerical values. They enable us to visually compare data records from distinct places and time periods, and simultaneous records for different variables when we wish to detect significant similarities or differences. What they do not do is give us numerical measures of central values, ranges, variabilities, etc., such as those shown in Table 3.2.

The latter are summary measures that report principal features of each data set and provide bases for quantitative comparison.

"Summary statistics" focus on characteristics which we may think of as providing "data profiles". These are numbers computed from a data set, which quantify where the data set is centered, the range of its values, how much variability there is among the values in the data set, and the approximate configuration of their cumulative frequency distribution. A single MINITAB command provides these statistics for each designated column of data. As noted in Section 3.1, DESCRIBE C2-C8 produced the output shown in Table 3.2 from the data in Table 3.1. The definitions and computational formulas of the statistics returned by the DESCRIBE command are reviewed in this section. You may wish to apply the formulas given here to the data of Table 3.1 and use the MINITAB output of Table 3.2 for confirmation, to familiarize yourself with any that are new to you. With MINITAB's notation, definitions of the statistics in Table 3.2 are as follows.

N is the total number of valid data values in the designated column, i.e., excluding missing data codes. MINITAB counts the number of missing data codes separately and reports this as N^*.

MEAN is the arithmetic average of the valid data values. In mathematical notation, $\bar{X} = \Sigma_j X_j / N$.

MEDIAN is the value which has as many data values less than it as there are greater than it. This is obtained by numerically ordering the data and identifying the middle datum if N is odd, or the average of the $N\backslash 2$ and $(N\backslash 2 + 1)$st of the ordered data values if N is even.

TRMEAN is computed by first removing the smallest 5% of the values in the numerically ordered data set and the largest 5%, and computing the arithmetic average of the remaining values. The notation is short for "trimmed mean". Using it in place of the mean is the same as excluding the largest and smallest values from consideration just because they are the largest and smallest: a practice of dubious logic.

STDEV is a measure of spread of the values in the data set about the MEAN. This is computed by first summing the squares of the differences between the individual observations and the MEAN, dividing the sum by $(N-1)$, and taking the square root of the quotient. In mathematical notation, $S_X = \sqrt{\Sigma_j (X_j - \bar{X})^2 / (N-1)}$.

SEMEAN is an estimate of the variability of means of similar data sets. It is computed by dividing STDEV by the square root of N: $S_{\bar{X}} = S_X / \sqrt{N}$.

MIN is the smallest value in the data set.

MAX is the largest value in the data set.

The *range of the data set* is from MIN to MAX; and we say that the data set spans an interval of length (MAX − MIN).

Q1 is the first quartile, i.e., the value with 1/4 of the data set less than it and 3/4 of the data set greater.

Q3 is the third quartile, i.e., the value with 3/4 of the data set less than it and 1/4 of the data set greater.

When $(N + 1)/4$ and/or $3(N + 1)/4$ are not integers, the quartile values Q1 and/or Q3 are obtained by interpolation. To illustrate this suppose the data values are ordered from smallest to largest, with their ordered sequence denoted by $X_{(1)}, X_{(2)}, \ldots, X_{(N)}$. Now we write $(N + 1)/4 = M + f$, where M is an integer and f is a fraction between 0 and 1, and define the first quartile as

$$Q1 = X_{(M)} + f\left[X_{(M+1)} - X_{(M)}\right].$$

When $3(N + 1)/4$ is not an integer, we write $3(N + 1)/4 = M + f$ and define the third quartile as

$$Q3 = X_{(M)} + f\left[X_{(M+1)} - X_{(M)}\right].$$

Note that when they are taken together, Q1, MEDIAN, and Q3 divide the data set into quarters. MEDIAN is Q2; that is, the value with 2/4 less and 2/4 greater than it.

With the exceptions of MIN and MAX, the values of the descriptive statistics need not be values actually in the data set. They are descriptives of the *distribution* of observation reports, which may not correspond to any of the recorded values.

Having met and considered several methods of summarizing the numerical values of a data set, you should now recall the detail of the time sequential data plot. Look at Fig. 3.1 again. The time series shown there provide complete descriptions of the records of their variables. What is lost in the summarizations which we have considered so far is all information about temporal trends in data.

A trend, or time coherency, is likely to be a strong characteristic of observations of atmosphere and ocean variables. We will consider characterization of this more challenging feature further on, when we introduce techniques for describing and analyzing both temporal and spatial trends.

3.4 COMPARISONS

Contrasting features of data records of one variable at different locations can be achieved with the same ease as assembling features of several variables at one location, provided the data have been stored in MINITAB

TABLE 3.3 Descriptive Statistics for January 1990 Daily Average Temperatures of the 15 Stations of the Data Set CLIMAT

	N	N*	MEAN	MEDIAN	TRMEAN	STDEV	SEMEAN
MRY	4	27	51.30	52.35	51.30	2.57	1.28
LAS	31	0	44.129	42.900	43.889	4.701	0.844
ABQ	31	0	33.26	32.20	33.19	5.58	1.00
FTW	31	0	51.44	51.90	51.24	6.92	1.24
NEW	31	0	56.59	56.80	56.50	6.48	1.16
CHS	31	0	53.49	54.30	53.77	6.29	1.13
MFR	31	0	38.255	38.000	38.067	4.382	0.787
BOI	31	0	33.66	33.90	33.49	6.96	1.25
LND	31	0	28.73	27.40	28.51	6.78	1.22
MKG	31	0	32.261	31.400	32.070	4.098	0.736
ALB	31	0	32.66	32.50	32.67	6.56	1.18
CAR	31	0	18.37	19.00	18.59	10.73	1.93
OME	31	0	2.82	4.30	3.21	14.97	2.69
FAI	31	0	− 13.40	− 12.30	− 13.86	16.82	3.02
ADQ	31	0	29.89	30.60	30.09	5.63	1.01

	MIN	MAX	Q1	Q3
MRY	47.50	53.00	48.62	52.92
LAS	36.900	56.400	40.600	48.000
ABQ	24.30	43.10	28.40	38.10
FTW	40.30	65.60	44.80	55.90
NEW	44.90	69.10	51.30	61.10
CHS	37.80	63.10	49.90	59.10
MFR	31.600	48.300	34.200	42.100
BOI	20.70	49.80	28.80	38.60
LND	15.60	45.20	23.40	32.90
MKG	24.000	43.000	29.400	35.000
ALB	19.70	50.00	28.00	37.30
CAR	− 2.60	36.50	11.70	26.30
OME	− 24.70	25.10	− 9.60	18.60
FAI	− 40.20	23.30	− 30.20	− 0.10
ADQ	19.20	37.80	26.20	34.80

file format. Thus, comparisons may be made of January 1990 daily average temperatures for geographically diverse locations in North America, with a DESCRIBE command. Table 3.3 illustrates this with station data for locations with widely different topography and latitudes.

3.5 CORRELATIONS

Thus far we have considered only descriptive features of sets of values of variables taken one at a time. Here, we define and discuss measures of *covariation* of values of variables in a data set, taken two at a time.

The *variance* of a single variable is defined as

$$V_X = \sum_j \left(X_j - \overline{X} \right)^2 / (n - 1).$$

This measures the degree to which the values in the collection of observations represented by $X_1, \ldots, X_n$ are dispersed around their arithmetic mean. Because the increments are squared, departures above and below the mean are weighted equally, so that it is a symmetric measure of dispersion. (You may recognize the variance as the square of MINITAB's STDEV.)

Let us now take two variables, represent their records by $X_1, \ldots, X_n$ and $Y_1, \ldots, Y_n$, where the indexing is over successive times at which values of both variables have been recorded, and consider the statistic

$$C_{X,Y} = \sum_j \left(X_j - \overline{X} \right)\left(Y_j - \overline{Y} \right) \Big/ (n - 1).$$

By contrast with the variance formulation, the terms of this quantity, and the sum itself, can be either positive or negative. If in the data set, values of Y above $\overline{Y}$ have been recorded on the same days as values of X above $\overline{X}$, and values of Y below $\overline{Y}$ have been recorded when values of X below $\overline{X}$ were recorded, then the components of each term in the summation have the same sign. Hence all summands, and the sum itself, are positive. On the other hand, if values of Y above $\overline{Y}$ were recorded for just those days when values of X were below $\overline{X}$, and values of Y below $\overline{Y}$ were recorded when values of X were above $\overline{X}$, the components of each term have different signs; and the result is negative. As a simple numerical example, consider the "data" in Tables 3.4(a) and 3.4(b), shown plotted in Figs. 3.6(a) and 3.6(b).

The *relationships* between the X and Y values plotted in Figs. 3.6(a) and 3.6(b) are of nearly identical strength. For Fig. 3.6(a), $C_{X,Y} = 0.993$,

TABLE 3.4   (a) Virtual Potential Temperatures with Height and (b) Specific Humidity with Height, over Ice of the Barents Sea

(a)	X	Y	(b)	X	Y
	0.04	2.40		0.04	4.48
	0.06	2.47		0.06	4.41
	0.08	2.95		0.08	3.93
	0.10	3.02		0.10	3.86
	0.14	3.56		0.14	3.32
	0.17	3.68		0.17	3.20
	0.23	4.44		0.23	2.44
	0.27	5.00		0.27	1.88

Adapted from Shaw, et al. (1991).

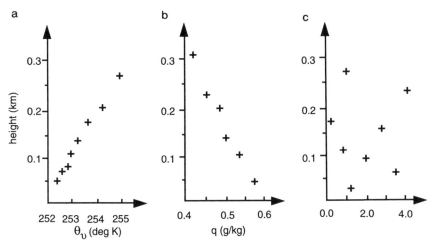

FIGURE 3.6 Plots of mean increment pairs for variables showing: (a) negative correlation, (b) positive correlation, (c) no apparent linear correlation.

while for 3.6(b), $C_{X,Y} = -0.992$. The magnitude of this measure, which we call *the covariance of X and Y* tells us how much the values vary together. The sign tells us whether they tend to increase and decrease together or inversely. Figure 3.6(c) presents a marked contrast with the nearly linear relationships of 3.6(a) and 3.6(b). In this third case there is no evidence of a linear (or nonlinear) relationship between the two variables.

The above discussion and example are a bit oversimplified. In most real situations the covariation of the values of two variables in a data set will show component product terms of generally consistent signs, with a few exceptions. The more consistent the signs and the magnitudes of the elements of the products, the stronger the covariation. Lack of covariation will show up in the absence of consistency: large and small increments with positive and negative signs will be mixed. The resulting summation will be somewhere in the vicinity of zero.

Ordinarily the covariance is not used by itself in the exploration of the relationship between two variables, because it confounds variability due to their relationship with their individual, inherent variabilities. The measure generally used to describe whether and how the mean increments of one variable increase and decrease linearly with those of another is the *correlation coefficient*. This is defined as the ratio of the covariance to the square root of the product of the variances of the values of X and Y. Thus it is equivalent to computing the covariance of "normalized" values of the data set: obtain this by dividing each mean increment by the standard deviation of the recorded values for that variable, as shown by the expression just below the defining equation for the correlation coefficient:

$$R_{X,Y} = \frac{\sum_j \left(X_j - \bar{X}\right)\left(Y_j - \bar{Y}\right)}{\sqrt{\sum_k \left(X_k - \bar{X}\right)^2 \sum_k \left(Y_k - \bar{Y}\right)^2}}$$

$$= \sum_j \left[\frac{\left(X_j - \bar{X}\right)}{\sqrt{\sum_k \left(X_k - \bar{X}\right)^2}}\right]\left[\frac{\left(Y_j - \bar{Y}\right)}{\sqrt{\sum_k \left(Y_k - \bar{Y}\right)^2}}\right].$$

Because the components of the summands have numerators and denominators in the same units, $R_{X,Y}$ is a unitless measure of linear association. If corresponding X and Y increments were identical, its value would be $+1$; if they were equal in magnitude with opposite signs, $R_{X,Y}$ would equal -1. For the recorded values of *any* pair of variables, the correlation coefficient will lie within $[-1, +1]$.[1]

Except for a totally trivial or redundant data set in which the recorded values for one of the variables are a constant, both sums of squares of mean increments will be positive. Without further qualifications in the following and throughout the text, it will be assumed that the values we deal with exhibit variability, both conceptually and in any set of recorded values. Thus the computed variance terms are positive and the correlation coefficient takes the sign of the covariance. If the mean increments of the variables have no linear relationship, the computed value of $R_{X,Y}$ will reflect this by taking a value in the vicinity of zero. In general, the stronger the linear relationship between the increments of the two variables, the larger the *magnitude* of the correlation coefficient; while the weaker the linear relationship between the increments, the closer R will be to zero.

The computations which yield values of correlation coefficients for variables in a data set may be done with a simple MINITAB command, e.g., CORRELATE C1 C2 with the values of Table 3.4(a) returned $R_{X,Y} = 0.993$.

3.6 WORDS OF CAUTION IN INTERPRETING VALUES OF DESCRIPTIVE STATISTICS

We have introduced techniques for describing and displaying characteristics of data sets. They give us several options for summarizing information *in an observation record* which we can use as tools of discovery and

---

[1]This follows from Schwarz' inequality: $[\sum_j (X_j - \bar{X})(Y_j - \bar{Y})]^2 \leq [\sum_k (X_k - \bar{X})^2][\sum_k (Y_k - \bar{Y})^2]$.

description. The output from TSPLOT illustrates the temporal coherence of the recorded values of the variables to which it is applied. HISTOGRAM and DESCRIBE permit us to see and quantify the frequency distribution of values for any of the variables in the data set, ignoring the time sequence of their numerical records. CORRELATE returns a scaled measure of the consistency with which recorded values for two variables deviate from their means, as a measure of a linear relationship between their sequences. *The focus here is on the data itself* in distinction to values of the same variables in the larger context of the system from which their observations were drawn and in distinction to any theoretical distribution properties.

The descriptive statistics have technical definitions, but it is important to interpret these only literally, in the absence of further knowledge of the statistical structure of the system from which the data records have been assembled. This cautionary note is particularly important to the interpretation of statistical summaries of data in the atmospheric and ocean sciences because of the strong spatial and temporal coherencies of the variable fields from which data are drawn. The classic associations of some of the statistics with properties of distributions with which you are familiar from other contexts, may not be valid in our data sets. Making automatic assumptions that they pertain here can restrict the potential scope of a data analysis and lead to conclusions which do not reflect all that the data can tell us. To extract all the information about the parent system, available in a set of observation records, it is important to approach the data without prior expectations.

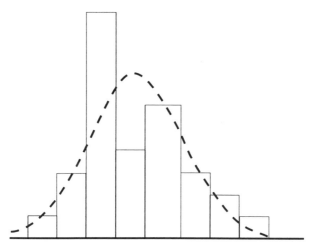

FIGURE 3.7 Histogram for October 1990 RELHUM at State College, scaled to correspond to the normal distribution with the same mean and variance, shown overdrawn.

We will return to this discussion again in following chapters. It is central to the challenge of making statistically based inferences with data records of the variables that index globally circulating fields. We conclude the present discussion with a simple illustration comparing an observed data distribution with the familiar "normal distribution curve". Figure 3.7 presents a histogram for the daily relative humidity values reported for State College in October 1990, from Table 3.1, overdrawn with a normal distribution curve having the same mean and standard deviation as the data. Comparison of the two makes it clear that even the best-fitting normal distribution is not a good approximation to the distribution for this set of observations and cannot be used to accurately characterize them. A normal distribution is always unimodal and symmetric, whereas the distribution of the humidity values is clearly bimodal and asymmetric. We should identify a basis for characterization, and subsequent inference, which more closely represents the true distribution structure of the observed variable. Our point here is that, before we apply a theoretical distribution in the context of a data set, we must first confirm that it gives a good approximation to what we have observed. Fortunately the tools we have available make this easy to do.

EXERCISES

1. Use the 94YEARS data set.
   (a) If you have not already done so, create a new column of monthly average temperatures: (MIN + MAX)/2, for all 94 × 12 months of the data set; and replace the original data set with this augmented file.
   (b) Create for new columns: (i) years and January values, for 1907–1946 and (ii) years and January values, for 1947–1986. Hint: Employ copy commands with qualifying subcommands on dates.
   (c) Make separate plots of January average temperature vs year, for the two subsets you created in part (b).
   (d) Repeat (b) and (c), for July average temperatures.
   (e) Make a new file containing just the columns you have created in this exercise, name the columns for future reference, and save this file.

2. Use the CLIMAT data set.
   (a) Create a new data set with the dates plus 14 columns containing surface pressures for all stations except Monterey (which didn't report surface pressure). Label the columns and save this new data set.
   (b) Create a "row number" column, numbering from 1 to 380.

(c) Plot daily surface pressures for Boise versus row number, from 1 to 365. Label the output by hand, marking off the four seasons. Clearly note the starting and ending dates in the figure caption.

(d) Repeat (c) with the daily surface pressures for Charlotte and Nome. If the plots for Boise, Charlotte, and Nome have different vertical scales, replot them so that they are comparable, i.e., so that they have the same scales.

3. Use the file you created in Exercise 1, containing 40-year subfiles of January and July average temperatures.

(a) Make four histograms with the HISTOGRAM command: (i) January 1907–1946, (ii) January 1947–1986, (iii) July 1907–1946, and (iv) July 1947–1986.

(b) If these are not comparable, by having the same interval lengths, redo them. Give them all the same interval lengths. Also give the same starting values to both January histograms and the same starting values to both July histograms.

(c) Comment on any interesting differences between the two time periods suggested by these histograms. How might you confirm these?

4. Use 365 consecutive daily surface pressure values for Boise, Charlotte, and Nome. Create *comparable* relative frequency distributions for these three locations, separately for the four seasons: fall, winter, spring, and summer. You will have four sets of three comparable relative frequency histograms when you are finished. Label them clearly, by hand. If you can't get them printed consecutively, use scissors and tape to lay them out on paper, for easy visual comparison.

5. Use the file you created in Exercise 1, containing 40-year subfiles of January and July average temperatures.

(a) If necessary, rearrange columns, with COPY/DELETE/COPY commands, so that the subsets of January values are in adjacent columns and the subsets of July values are adjacent.

(b) Use DESCRIBE to obtain the descriptive statistics for these four columns.

(c) Pick out and copy by hand the values which divide these subsets into quartiles. (Comment on any differences you notice between these descriptive statistics for the two 40-year periods.)

(d) For each of the four subsets, write down the range of temperature values within two standard deviations of the observed mean. Label these clearly, so that you will recognize them six weeks from now!

6. Use the 40-year average, January and July, temperature file.

(a) Create new columns, with 10 values in each. These 10 values will be for consecutive decades: 1907–1916, 1917–1926, ..., 1977–1986. Do this first for January and then for July.

(b) DESCRIBE the columns for the eight consecutive decades of January values.

(c) Pick out the median values for each decade.

(d) Create two new columns: one with consecutive integers $1, \ldots, 8$, and the other with the median values from part (c).

(e) Plot the decadal median values versus the integers which index the decades, and clearly label your output.

7. Use the 94YEARS data set.

(a) For each of the 12 months, create a column containing the values for the most recent 25 years. (Hint: By referring to your *MINITAB Handbook*, you should be able to do this with a single conditional COPY command.)

(b) Obtain the descriptive statistics for the 12 columns you have just created.

(c) With the SET command, create 3 new columns: the first containing the consecutive integers $1, \ldots, 12$, the second with the mean values from the output of step (b), and the third with the standard deviation values from the output of step (b).

(d) Make two plots: one of the mean values versus the integers and the second of the standard deviation values versus the integers. Label these clearly, by hand, identifying the months with their integer indices, and the specific years included in the averaging.

8. Use the STN#27 data set.

(a) For the temperature subset, DESCRIBE the columns corresponding to each of the 12-month means.

(b) Pick out the range value for each of the 12 months and plot these versus month.

(c) Repeat (a) and (b) for the salinity subset.

(d) Create new subsets which reverse depth and month. This will give you 10 columns of 12 temperature values, one column for each depth; and 10 columns of 12 salinity values. Add an additional column whose elements are the integers $1, \ldots, 12$, that index the months. Make these new columns by whatever technique is easiest for you; and this may be simply using the SET command and typing them in, from a paper copy of the data. Name the columns so that you will recognize them later, and save this new data set.

(e) Plot each of the 10 columns of salinity data versus month index, using comparable vertical scales. Use the MINITAB label command. Now replot the columns corresponding to the different depths all in the same frame, using MINITAB's multiple plot command.

4

THE FOUNDATIONS OF INFERENCE
"probability models as descriptions of research outcomes"

4.1 INTRODUCTION

We have reviewed descriptive statistics in the previous chapter. Collectively they provide a mechanism for summarizing the composition of a data set. Provided the data set is extensive and provided the techniques used to assemble it permit us to treat it as representative, the statistics which describe this data set may be presumed to describe characteristics and relationships among values of the observed variables *in a larger context*. Specifically, they may be thought of as describing characteristics of, and relationships among the observed variables over the entire region and period of time from which the observations were obtained. Given that we heed the tenets of representativeness, the more information we have about the system being observed, the more confident we are that we can truly characterize it.

The extensiveness of a data set and the degree to which we may presume that it represents a greater reality are relative to the physical extent and complexity of the system being observed. As a general rule, in the contexts of oceanic and atmospheric research, the extent and complexity of a system will limit confidence that the data alone are a sufficient representation. Generally we couple the evidence of present observations with assumptions about the "statistical structure" of the system, derived from earlier observation records and from scientific theory. These assumptions provide a format for making inferences from observations to the larger context of the field variables. These assumptions are phrased as "probability models". And they may be used most effectively when their derivations can be seen clearly in relation to the objectives, protocol, and procedures for a research program. We now take on the challenge of establishing this connection.

At first reading, the material of the following sections may seem like an excursion of mathematical fancy into regions foreign to science. However, it will lead to an understanding of just how critical the validity of assumptions about statistical structure is to the validity of scientific inferences. Accordingly, it is the foundation for confidence in the creation, deployment, and interpretation of valid statistical procedures. In this chapter we develop essential tools for using the framework of estimation and hypothesis testing as an effective component of research design and analysis. We ask you, Eminent Reader, to approach this undertaking with a spirit of adventure, curiosity, and faith in a powerful outcome.

4.2 THE BASIC PROBABILITY MODEL

The probability model is defined by the statement of the research objective and the physical/technical circumstances of the scientific program it serves. These determining factors include the synoptic state of the field being studied, instruments used for sensing and recording field variables, and the timing and spatial array of the observing protocol. Together they implicitly define all of the following three elements of a probability model:

    (i) the set of possible values for the variables being observed,
    (ii) the events for which probabilities can be assessed,
    (iii) the rule for assigning probabilities to the events.

A simple illustration will support the introduction of basic definitions and concepts of probability models. Let's say that we state our objective to be the prediction of the number of days on which there will be significant precipitation as snow in State College, PA,[1] next January. With 31 days in January, the range of possible values for our enumeration variable is from 0 to 31. The formal name we will give to the inclusive set of possible values

---

[1]State College is a location for which there are good instrumentation, reliable observations, and a long record on which to base expectations.

is *the description space* and we will designate it with the letter S. Here

$$S = \{0, 1, 2, \ldots, 31\}.$$

Clearly the values in the description space have widely differing likelihoods. From the State College precipitation records, we know that it is extremely unlikely that it snows every day of the month. In fact it would set a new record for the century if significant snow fell on as many as 10 days in January. Nonetheless we cannot say with absolute confidence that it will not snow on 10 or any greater number of days. So we must include all possibilities in S, without regard for their relative likelihoods.

An *event* is technically defined as "a subset of S". However, we will generally designate a specific event in terms of a happening as it is described in English. In the context of snow in State College in January, "significant snowfalls on more than five days of the month" is an event. We can pick out the values in S which correspond to this event, write it as

$$E = \{6, 7, 8, \ldots, 31\},$$

and see that it fits the technical definition. "No significant snowfalls in January" is written as {0}. Exercise 1 will give you practice in translating from verbal descriptions of outcomes to designations of events as subsets of the full, outcome description space.

The collection of all events for which probabilities may be assessed is denoted by **E**. In the present example, or any other example for which the outcome description space is finite or countable, we can designate **E** to be the collection of all subsets of S. In every case, **E** will be understood to include *the empty set* $\varnothing = \{ \ \}$, which is the set containing no value, and the set of all possible values, which is S itself.

To indicate that an observable value s belongs to an event E, we write

$$s \in E;$$

and we say *the event E occurs* whenever the value of the observed variable is one of the values in the subset defining E.

For a countable description space, we may assign non-negative probabilities to every value in S and assess the probability of any event E occurring by summing over those values in E:

$$P[E] = \sum_{s \in E} P[s]. \tag{4.1}$$

Since we have postulated that the values of S describe all possible outcomes, then S is *the certain event* and it must have probability 1. Additional criteria for probability assignments are that the empty set, or *null event*, has probability 0, and that the probability of every event in **E** will be non-negative and bounded by 1. We write these as

$$P[S] = 1, \qquad P[\varnothing] = 0, \qquad \text{and} \qquad 0 \le P[E] \le 1, \forall E \in \mathbf{E}. \tag{4.2}$$

We may wish to associate probabilities with events of the future, by using a good historical record. Thus we may choose to assign the frequencies with which the events have occurred in the past. In our current example, we might interrogate the whole 94-year weather record for State College, and assign as the probability for any event E the proportion of those 94 years in which the event occurred. For example, there were 57 years in which there was significant snowfall on more than 5 days and 0 years in which there was no significant snowfall during January. So we would assign

$$P[\{6,7,8,\ldots,31\}] = 0.6 \quad \text{and} \quad P[\{0\}] = 0.0.$$

If there is reason to believe that there has been a microclimatic change during recent decades, then we should not include the whole of the 94-year record for the assignment of probabilities pertaining to next January, but only those years which we believe represent the current regime. (Historical records must be used with scientific discretion.) See Exercises 2 and 3 for probability estimates based on a recent 10-year period.

There are many other ways of establishing rules for associating probabilities with events; and we will meet others in other contexts. Frequently research is carried out for the purpose of exploring a hypothesis for which there is no prior observation record. The research proposal specifies both the variables which will be directly or indirectly observed and the mechanisms for obtaining observations, including error characteristics of the mechanisms for sensing, transmitting, and recording data. These specifications, together with an exploratory hypothesis, form the basis for stating a probability rule. If they do not do this completely then the research proposal lacks information critical to quantitative assessment of the relative likelihoods of research outcomes. Without complete specifications we lack the tools to assess the significance of research results; and some additional groundwork is required. To determine whether a research proposal is complete in this respect, go back to the three elements of the basic probability model and ask whether the proposal defines them, either implicitly or explicitly. If it does not, rewrite it so that it does. To see how this works, use the scientific question you developed for Chapter 1. The acid test is whether you can asess the probabilities of the outcomes on which the answer to your question pivots.

4.3 CONTINUOUS OUTCOME DESCRIPTION SPACES AND EVENT SET OPERATIONS

Most variables we will deal with are continuous in nature, rather than having only a finite set of values. Temperature, humidity, salinity, wind,

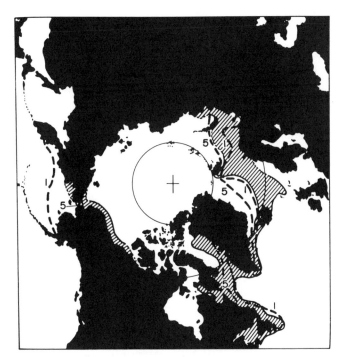

FIGURE 4.1 The Arctic basin, with the biologically productive region shaded, and the 1/10 and 5/10 ice concentration contours labeled in tenths (from Mysak and Manak, 1989).

and current speeds are examples.[2] For any one of these, the set of all possible values will be a continuous interval, such as $S = [0, 100]$ for humidity or salinity as a percentage of saturation. Variables like temperature and speed have absolute lower limits but no specific upper limits. Clearly there are values above which temperature and speeds have never been recorded. However, because we cannot name the least large value that will ever be reported, we would write $[-273, +\infty)$ and $[0, +\infty)$ for temperature and speed, for example. This simply means that we are not naming a limit on the right, in either case. The rule for associating probabilities with subsets of S can handle inclusion of impossible values rather neatly by attaching zero probabilities to them. Accordingly, for any continuous variable, one option is to write $S = (-\infty, +\infty)$, which is the whole real line, and handle practical constraints with judicious assignment of probabilities.

---

[2]Although measurement accuracy and convention limit the number of decimal places recorded, it is easier to treat these variables as if their reported values also have continuous ranges, for the modeling and analysis of their distributions.

As a specific illustration, consider Arctic sea ice extent, as it is reported by Mysak and Manak (1989) for example. This is a variable which is measured in units of areal extent, km^2. At any one observation time, the value must be a number somewhere in the range from 0 to the total ocean area of the Arctic basin (see Fig. 4.1). We may either define S to be the interval from 0 to the total area T, or we may let $S = (-\infty, +\infty)$ and assign zero probabilities to the intervals $(-\infty, 0)$ and $(T, +\infty)$ which contain all the numbers outside the range of what we might observe. The latter convention is useful because it provides a single numerical description which covers all possible scalar variables. We will make use of this in the development of the generic probability model.

When the description space is a continuous interval, characterization of the collection of events for which we can evaluate probabilities requires special attention and thought. It is a mathematical impossibility for all values in the interval to have positive probability,[3] so the subsets defined by single values of S are not admissible elements of **E**. To define **E** we start with all the half-open subsets of S, of which the interval $[0, 1.5 \times 10^6$ km$^2)$, including zero and all greater numbers less than 1.5×10^6 km^2, is an example. Then join with these the complements, intersections, and unions of all pairs of members, as they are defined below. Together these subsets designate every event for which we might reasonably seek to evaluate a probability, including S and $\varnothing$. The definitions we need are the following.

For any event E, *the complement of E* is the set of values in S that are not in E, which is written as

$$E^c = \{s \in S: s \notin E\}. \tag{4.3}$$

E^c *occurs* whenever E does not occur.

For any pair of events A and B, *the intersection of A and B* is the set of values these two events have in common. This is denoted by

$$A \cap B = \{s \in S: s \in A \text{ and } s \in B\}. \tag{4.4}$$

$A \cap B$ *occurs* whenever both A and B occur.

The union of events A and B is the set of values in at least one of A and B. This is written as

$$A \cup B = \{s \in S: s \in A \text{ or/and } s \in B\}. \tag{4.5}$$

$A \cup B$ *occurs* whenever A or B, or possibly both, occurs.

---

[3]The rigorous proof of this will not be given here. We simply state the logical argument that if each of an uncountable infinity of points had positive probability, their collective probability would greatly exceed 1.

Using humidity/salinity records for illustration, the three set operations for events may be diagrammed as

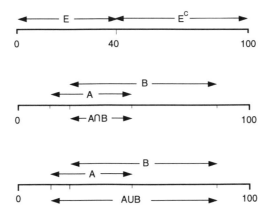

In addition to the definitions above, it will be useful to have the following.

A not B, denoted by $A - B$, is the set of values in A that are not also in B. $A - B$ *occurs* whenever A but not B has occurred.

$B \subset A$ means that every possible outcome in B is also in A. In English: the occurrence of B implies the occurrence of A.

We note some immediate consequences of the definitions which will aid in evaluating probabilities in our later work.

For any event $E \in \mathbf{E}$,

$$E \cup E^c = S, \quad \text{where } E \cap E^c = \varnothing, \tag{4.6}$$

i.e., E and E^c make up the complete set of possibilities, and they are mutually exclusive.

$$A = (A \cap B) \cup (A - B), \quad \text{where } (A \cap B) \cap (A - B) = \varnothing, \tag{4.7a}$$

i.e., for any event B, A can be written as the union of mutually exclusive subsets: those outcomes that correspond to the occurrence of B and those that correspond to B not occurring. Since $A - B = A \cap B^c$, the above can also be written

$$A = (A \cap B) \cup (A \cap B^c), \quad \text{where } (A \cap B) \cap (A \cap B^c) = \varnothing, \tag{4.7b}$$

$$B \subset A \text{ implies } A \cap B = B, \tag{4.8}$$

$$A \cap B = \varnothing \text{ implies } B \subset A^c. \tag{4.9}$$

For continuous description spaces for which single point sets are not admissible elements of **E**, the rule for assigning probabilities takes the form of a *probability density function* (p.d.f.), f. The way this works is that the probability of any event E is given by the integral of $f(s)$ over all of the possible outcomes corresponding to occurrences of E:

$$P[E] = \int_E f(s)\, ds. \qquad (4.10)$$

Since the criteria for probabilities summarized by (4.2) must hold for continuous description spaces as well, we have the following conditions on the p.d.f.

$$\int_S f(s)\, ds = 1, \qquad \int_\varnothing f(s)\, ds = 0, \qquad \text{and} \qquad 0 \le \int_E f(s)\, ds \le 1, \forall E \in \mathbf{E}. \qquad (4.11)$$

f must be a bounded, non-negative function with a value at every point of S. Generally the probability density function (p.d.f.) will also be continuous. If we have adopted the option of writing $S = (-\infty, +\infty)$, the p.d.f. may be zero on all but a finite interval of the real line, which we must take care to specify.

Equation (4.10) is the continuous analogue of a probability summation for a discrete description space (4.1). However, it is vital to note and remember that $f(s)$ is not a probability. *It is the function whose integral over any region of the description space gives us the probability of the research outcome being in that region.* Any one real number or point $s \in S$ has probability 0.[4]

4.4 PROBABILITY MODELS FOR VECTOR-VALUED OUTCOMES

Here we expand the probability model concept to include more than one variable at a time. This might be as simple as designating a pair of variables, say daily minimum and maximum temperatures, or as complex as the simultaneous treatment of all the variables from the global observing network which are used in numerical weather prediction. Regardless of the number of variables, the points of the description space will be vectors, each representing a possible combination of observable values for these variables. The *dimension of S* is the number of scalar variables being considered together, which will then also be the length of each vector

---

[4]This may seem oddly inconsistent with the fact that recorded values *are* single point values and they *do* represent true states of the field. However, because of the limits of measurement devices and the truncations of values entering data records, a recorded value represents a (small) continuum of possible true values and thus has an assignable positive probability. In developing a probability model it is important to distinguish between the values available to a field variable and its measurements records.

describing a possible outcome. We note that in the initial discussion of an outcome description space, dimensionality was not an issue. The extension to general dimension is immediate, with the interpretation that elements of S are vector valued. Designation of the composition of the collection of assignable events, **E**, and specification of the rule for associating probabilities with the events of **E** will again be determined by whether S is discrete or continuous. If it is discrete, then again we may let **E** be the collection of *all* subsets $E \subset S$, assign a probability to every point $s \in S$, and evaluate the probability of any event E by summing the individual probabilities of its elements:

$$P[E] = \sum_{s \in E} p(\mathbf{s}).$$

If S is a continuous region of Euclidean m space, which we will write as R^m, then the definitions of the first two elements of the probability model require a bit more work, just as they did in one dimension.

To specify the collection of events for which probabilities may be evaluated and then the requirements which must be satisfied by a rule which assigns probabilities to them, in the continuous, multidimensional context, we begin with an m-dimensional analogue of a half-open interval. For example, with (TMIN, TMAX) as the vector

$$\{(t_{\min}, t_{\max}) : t_{\min} \le a \quad \text{and} \quad t_{\max} \le b\}$$

qualifies as a *half-open region of S*. We now define **E** to be the collection of all half-open regions of S, together with their complements, and the intersections and unions of all pairs of these regions. The rule for assigning probabilities is again in the form of a probability density function (p.d.f.), although in this case it is defined in R^m and the probability of any $E \in \mathbf{E}$ is evaluated as

$$P[E] = \int \cdots \int_{\{(s_1, \ldots, s_m) \in E\}} f(s_1, \ldots, s_m)\, ds_1 \ldots ds_m.$$

Another way of writing this is

$$P[E] = \int_E f(\mathbf{s})\, d\mathbf{s}. \tag{4.13}$$

Again we require

$$P[S] = 1, \qquad P[\varnothing] = 0, \qquad \text{and} \qquad 0 \le P[E] \le 1, \forall E \in \mathbf{E}$$

which imply that f is a non-negative bounded function defined on R^m, that integrates to 1 over the whole space.

Illustration with a Continuous, Two-Dimensional Probability Model

The example of daily minimum and maximum temperatures is a good one to illustrate the general formulation of the probability model. Expected values for the pair (TMIN, TMAX) provide guidelines in planning for the use of time and resources in agriculture, recreation, and travel. Evaluating the probabilities that this pair of variables are within specified limits is highly germane. To achieve this objective we first construct a probability model from our store of scientific information.

For the vector variable (TMIN, TMAX) we can either define the description space, S, to be the set of all points of R^2 which are possible values for the pair, or we can designate $S = R^2$ and let the probability rule handle physical constraints by appropriate assignment of zero probability to unrealizable combinations. It is generally a good practice to do the former in the process of defining the probability model, to avoid physically nonsensical regions, and then relax the designation to the whole space after the probability rule has been correctly specified. In our example, we can say with confidence that the components of (TMIN, TMAX) for surface air temperatures on our planet will be between $-100°$ and $+200°$F, with TMIN $\leq$ TMAX; and we can diagram S as shown below.

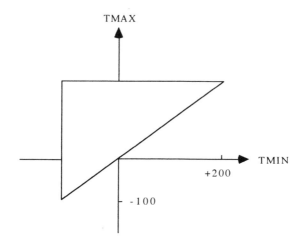

Events may be defined by limit specifications, such as

$$E_1 = \{(t_{min}, t_{max}): t_{min} < 15 \text{ and } t_{max} \geq 25\}$$

$$E_2 = \{(t_{min}, t_{max}): (t_{max} - t_{min}) = 5\}$$

$$E_3 = \{(t_{min}, t_{max}): t_{min} \geq 32\}$$

$$E_4 = \{(t_{min}, t_{max}): t_{max} \leq 85\}$$

$$E_5 = \{(40.2, 48.6)\},$$

and drawn as subregions of S. The first is diagrammed as shown:

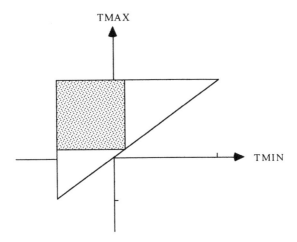

The rest are left as exercises.

In the context of the current example, guidance in formulating a continuous probability rule may be obtained from temperature records. To illustrate, we will use daily minimum and maximum temperatures for Crookston, MN (CRKS), for May, a month in which these are critical to agriculture. We begin with the construction of a relative frequency histogram using the records for Crookston, in the data file TEMPS. From this we can estimate the p.d.f. for the temperature pair (TMIN, TMAX). Because we are working with two variables at once, this cannot be done directly with MINITAB. However, we can use the plot command PLOT C4 VS C3 and construct the frequencies for intervals of daily (TMIN, TMAX) values by hand, by counting the numbers of values within equal-sized boxes. Exercises 10 and 11 provide detailed guidance in carrying this through to a contour plot of a relative frequency surface. For the TEMPS data set, the intervals are in $(°F)^2$. Since each relative frequency, say h, is the proportion of values in the data set, in the corresponding $(\Delta t) \times (\Delta t)$ interval, the desired probability density function is one which fits the set of $h(t^*_{min}, t^*_{max})/(\Delta t)^2$ values reasonably well, where the (t^*_{min}, t^*_{max}) are interval center points. The choice of functions generally considered to represent probability densities and the methods for matching them with observed histograms are topics addressed in later chapters. We defer further work on this until we are better equipped technically. In general, we will be seeking a density function f for which

$$f(s) = h(s^*)/|\Delta s|,$$

where h is from the relative frequency file constructed for a histogram with base intervals of size $|\Delta s|$ centered at the s^* values.

4.5 GENERIC EVENTS AND PROBABILITY RULES

"Venn Diagrams" are classical aides to learning the rules for evaluating probabilities. They depict S and subsets of S diagrammatically as regions of the plane, as shown here:

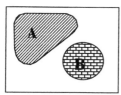 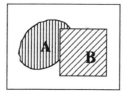 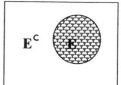

These sketches provide a generic rendition of all the relationships between events that we are likely to find useful in scientific inference.[5] We will use them here for reference in introducing rules for calculating probabilities.

Recall the requirements for probability rules:

$$P[S] = 1, \qquad P[\varnothing] = 0, \qquad \text{and} \qquad 0 \le P[E] \le 1, \forall E \in \mathbf{E}.$$

Also recall that the specific rules for assigning probabilities to events take different forms for discrete and continuous description spaces. When S is discrete

$$P[E] = \sum_{s \in E} p(s),$$

where $p(s)$ is the probability assigned to the single point set $\{s\}$. When S is continuous

$$P[E] = \int_E f(s)\, ds,$$

where f is a probability density function.

---

[5]A good discussion of Venn Diagrams in the more formal context of statistics is given by McClave and Dietrich, II (1991).

From the tenets of the previous paragraph we can derive all the rules we will need in formulating the structure of statistical inference in science. In probability language, the keys are the following.

For *disjoint* events, A and B of $\mathbf{E}$, $P[A \cup B] = P[A] + P[B]$.   (4.14)

For *any* events A and B, $P[A \cup B] = P[A] + P[B] - P[A \cap B]$.

(4.15)

For any event E, $P[E^c] = 1 - P[E]$.   (4.16)

The proofs of these assertions for discrete description spaces are given below. You can easily rephrase them for continuous description spaces.

Proof of (4.14).   For disjoint A and B, there are no points in both A and B. Thus

$$P[A \cup B] = \sum_{s \in A \cup B} p(s) = \sum_{s \in A} p(s) + \sum_{s \in B} p(s) = P[A] + P[B];$$

i.e., we include each value in $A \cup B$ exactly once, in summing first over the values of A and then over those of B.   ∎

Proof of (4.15).   To evaluate $P[A \cup B]$ when the events A and B may have values in common, we must factor $A \cup B$ into the union of disjoint events, and then use (4.14). We do this in two steps. First write

$B = (B - A) \cup (B \cap A)$,   where $(B - A)$ and $(B \cap A)$ are disjoint,

so that we can use

$$P[B] = P[B - A] + P[B \cap A]$$

and thus

$$P[B - A] = P[B] - P[B \cap A].$$ (4.17)

Now write

$$A \cup B = A \cup (B - A),$$

where A and $(B - A)$ are clearly disjoint, so that

$$P[A \cup B] = P[A] + P[B - A].$$

Finally we substitute the right-hand side of (4.17) for $P[B - A]$ to get

$$P[A \cup B] = P[A] + P[B] - P[B \cap A]$$

as claimed.   ∎

Proof of (4.16).   Since E and E^c are disjoint, by definition, and $S = E \cup E^c$ then

$$1 = P[S] = P[E] + P[E^c]$$

and hence

$$P[E^c] = 1 - P[E].$$ ∎

These three rules for evaluating probabilities for combinations of events, when we have some knowledge of their elements, will be used over and over in the work that follows.

4.6 CONDITIONAL PROBABILITY AND INDEPENDENT EVENTS

Conceptually the probability of an event A is the frequency of occurrence of this event among all the occasions on which it might occur. Another way of thinking about it is that, if the Weather Gremlin or the Ocean Troll had a very large number N of situations to select from, *each with the same likelihood of being selected*, then the probability of A is just $1/N$ times the number of those which correspond to "A occurs". If this number is k_A then $P[A] = k_A/N$. This is the probability of A if we have no "insider information" about the state of the atmosphere or ocean; and we call it the *unconditional probability*.

There are many ways in which we might create a special situation or have gained peripheral knowledge affecting the likelihood of event A. Generally this can be phrased in terms of the occurrence of another event, which we will call B. Say that we know B has occurred. Now the problem is to assess *the conditional probability that event A occurs, given that we know B has occurred*. We denote this as $P[A|B]$. We are no longer at large in S. Rather, we know that we are dealing with just those situations corresponding to B. Consider the Venn Diagram:

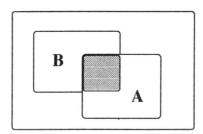

Since B has occurred, the only way for A to occur also is for the outcome to be in the intersection $A \cap B$. Since the reference set is now B, then k_B is the total number of situations from which the Gremlin or Troll is choosing. The proportion of those that correspond to A is $k_{A \cap B}$, so that

$$P[A|B] = k_{A \cap B}/k_B.$$

If we now divide the numerator and denominator by N we get

$$P[A|B] = (k_{A \cap B}/N)/(k_B/N) = P[A \cap B]/P[B].$$

In general, i.e., with or without the aid of gremlins and trolls, *the computational formula for the conditional probability of A given B* is

$$P[A|B] = P[A \cap B]/P[B]. \tag{4.18}$$

Of course this presumes that B is a possible event, with $P[B] > 0$, since otherwise the concept of probability conditioned on B does not make sense. (We exclude fictitious circumstances from consideration.)

The independence of two events is defined in terms of conditional probabilities: *A and B are independent* if

$$P[A|B] = P[A]. \tag{4.19}$$

Since (4.18) enables us to write $P[A \cap B] = P[A|B] \times P[B]$ for any pair of possible events, then if A and B are independent we may substitute $P[A]$ for $P[A|B]$ and write

$$P[A \cap B] = P[A] \times P[B]. \tag{4.20}$$

The latter is sometimes used as the defining condition for independence. However, its statement is less clear than the verbal reading of (4.19), namely: "The fact that B has occurred does not alter the probability of the occurrence of A." Nonetheless, (4.19) and (4.20) are equivalent. Both will be true if one is.

There are many, many uses of the concept of independence in statistical analysis; and it has a special relevance in ocean and atmospheric science. It is especially important to understand what it says, and what it does not say, about subsets of a description space for research results. Clearly, if two possible events A and B cannot both occur, i.e., $A \cap B = \varnothing$, then A and B are not independent because $P[A|B] = 0 \ne P[A]$. On the other hand, if A and B are two possible events for which $B \subset A$ without being identical to A, then A and B are not independent because $P[A|B] = 1 \ne P[A]$, unless $A = S$.

We will meet the concept of independence again in discussing random variables, where it plays a critical role in defining and using their distributions for inferences. In this context *independence* means that knowledge of the value of one variable does not influence the values available to others or the probabilities of their taking on those values.

Before leaving this section, we introduce a theorem from the theory of probability which has many useful applications in geophysical research. We write it here in terms of probabilities of events, although it may be written in different formats, depending upon the designation of the probability rule.

BAYES' THEOREM. *For any two possible events A and B,*

$$P[A|B]P[B] = P[B|A]P[A]. \qquad (4.21)$$

Proof of (4.21). Make two applications of the defining equation for conditional probability:

$$P[A|B] = P[A \cap B]/P[B] \quad \text{and} \quad P[B|A] = P[A \cap B]/P[A].$$

Multiply the first equation by $P[B]$ and the second by $P[A]$, and equate the left-hand sides. ∎

4.7 AN APPLICATION OF BAYES' THEOREM

We close this chapter with an illustration of Bayes' Theorem (E. S. Epstein, personal communication, 1991), with an assessment of the probability that a particular year will be classified as an "El Nino Year" (ENY). Information available well in advance of events that determine whether or not a year is classified as an El Nino Year permits early assessment of this probability. Both the interval of time since the most recent El Nino and the fall-to-winter sea surface temperature (SST) differential in a critical area of the Pacific are keys to this determination. Analysis of recorded intervals between El Nino events indicate that once at least 3 years have passed since the last event, the probability that the succeeding year will be an ENY is constant at 0.40. With 53 years of sea surface temperature data from the critical ocean area and the record of ENY classifications for those years, we know that, for the 12 which were classified as El Nino Years, the Oct/Nov to Jan/Feb temperature changes had a mean of $5.35°$ C and a standard deviation $0.623°C$. For the non-ENYs, these statistics are $4.28°C$ and $0.487°C$. For the year we have chosen as an example, the prior El Nino had begun more than 3 years earlier and the fall-to-winter temperature differential was $5.35°C$ (warmer in Jan/Feb than during the preceding Oct/Nov).

We will use two applications of Bayes' Theorem with the information available in March, to calculate the probability that this will be an El Nino Year. We designate the event of primary interest to be

A: "This will be classified as an El Nino Year."

and the conditioning event to be

$$B: \text{``}\overline{\Delta T} = 5.35°C.\text{''}$$

We use the information we have from analysis of El Nino records to evaluate the elements of the right-hand sides of the two expressions

$$P[A|B]P[B] = P[B|A]P[A]$$
$$P[A^c|B]P[B] = P[B|A^c]P[A^c]. \qquad (4.22)$$

The second of these is just Bayes' Theorem applied to A^c and B.

Finally, we form the ratio of the expressions in (4.22), from which $P[B]$ cancels; and apply the relationship

$$P[A^c|B] = 1 - P[A|B]$$

to reach our objective, namely a formula for the calculation of $P[A|B]$:

$$P[A|B]/\{1 - P[A|B]\} = \{P[B|A]P[A]\}/\{P[B|A^c]P[A^c]\}.$$

Since the interval between the year under consideration and the prior ENY is at least three years we know that

$$P[A] = 0.40 \quad \text{and} \quad P[A^c] = 1 - P[A] = 0.60.$$

The results of the analysis of the sea surface temperature differentials of the 53 prior years permit us to assign values to the conditional probabilities on the right-hand side of (4.22). Specifically, since the observed temperature differential coincides with the mean for previous ENYs and is more than two standard deviations from the mean differential for previous non El Nino Years, we may assign

$$P[B|A] = 1.0 \quad \text{and} \quad P[B|A^c] = 0.05.$$

Thus we have for (4.22),

$$P[A|B]P[B] = 1.0 \times 0.4$$

and

$$P[A^c|B]P[B] = 0.05 \times 0.6.$$

Their ratio is

$$P[A|B]/\{1 - P[A|B]\} = 40/3.$$

And the probability we seek, namely the probability that this will be an ENY, given the data available in March, is now calculated to be

$$P[A|B] = 0.93.$$

Our advance information has greatly increased our expectation that this will be classified as an El Nino Year, over the initial value of 0.40.

EXERCISES

1. For each of the variables defined below, state whether you can treat its outcome description space as continuous; and, if not, why not. Write the numerical description of S and of the designated event, in set notations.
   (a) Variable = the humidity at Washington National Airport at noon on August 7, next summer. Event: "The humidity exceeds 90%."
   (b) Variable = the number of storms that bring hurricane force winds to land at Charleston, SC, during the passages of the next five storms. Event: "Charleston escapes hurricane force winds."

(c) Variable = the distance from Sydney, Nova Scotia, to the nearest edge of the Gulf Stream. Event: "The Gulf Stream is within 125 km."

(d) Variable = the number of icebergs which come within 0.5 km of a research vessel during the month of July, when it is anchored on Fylla Bank. Event: "The vessel must be moved at least 8 times, if it is moved whenever an iceberg gets as close as 0.5 km."

(e) Variable = the possible area of an oil slick, 5 days after a spill in the Gulf of Mexico. Event: "The oil slick is successfully cleaned up within 3 days of the spill."

2. Use the 10-year State College data set to construct an estimate of the distribution of the number of days of significant snowfall we can expect in the coming January. That is, assign a probability to each one of the values in the set of all possible outcomes:

$$S = \{0, 1, \ldots, 31\}.$$

Do this using MINITAB commands which create 10 new columns: one for each of the 10 years; and then get information about these columns. Specifically, you use thrice-conditioned copy commands:

COPY C8, C10;	COPY C8, C11;	... COPY C8, C19;
USE C1 = 1980;	USE C1 = 1981;	... USE C1 = 1989;
USE C2 = 1;	USE C2 = 1;	... USE C2 = 1;
USE C8 = 0.1 : 100.0.	USE C8 = 0.1 : 100.0.	... USE C8 = 0.1 : 100.0.

Then use the INFO command on these new columns and read off the number of days of significant snowfall for each of the 10 years. The condition these instructions place on which C8 values MINITAB should copy specifies the range from the smallest value which is recorded for snowfall to a value we believe comfortably exceeds any that State College has experienced in a single day. (If you wish, you can replace the 100.0 value with the actual maximum amount of snowfall on a single day, during these 10 years, by using the DESCRIBE C8 command and finding this number in the output.)

The number of elements in each column is the number of days of significant snowfall in January of the corresponding year. So the collection of these 10 numbers are the values of the variable actually observed during the 10 years and, thus, are the numbers from which you are to construct an estimate of the distribution.

3. First confirm that the result of Exercise 2 satisfies the requirements of a probability rule as they are given in (4.2). [Does $\sum_{s \in S} p(s) = 1$?] Then find the probabilities of the following events, using this probability rule.

(a) Event: "There are no days of significant snowfall in January."

(b) Event: "There are either 4 or 5 days of significant snowfall."

(c) Event: "There are fewer than 3 days of significant snowfall."

(d) Event: "There is at least 1 day of significant snowfall."

4. Revisit the scientific question of Exercise 2 in Chapter 1. Focus on the one variable you think is most important in answering this question. For this variable, see if you can define or characterize the three elements of the basic probability model:
   (i) the set of possible values for this variable,
   (ii) the events for which probabilities can be assessed,
   (iii) a rule for assigning probabilities to the events of (ii).
   If the probability rule presents a problem, say in one or two sentences what you will need to know in order to define this element.

5. Refer to Exercise 1(a) above.
   (a) In a manner similar to illustration on page 59, diagram the events:
      A: "The humidity at Washington National Airport is at least 85%."
      B: "The humidity at Washington National Airport is at least 95%."
      and $C = A \cap B$.
   (b) Are A and B mutually exclusive ($A \cap B = \varnothing$)?
   (c) Does A imply B?
   (d) Is $B \subset A$?

6. Refer to Exercise 1(b) above.
   (a) Apply the notation that describes events as sets of possible outcome values to the number of Atlantic hurricanes that bring gale force winds to Charleston, SC; and describe the events:
      A: "All of the next five bring gale force winds to Charleston."
      B: "Two of the hurricanes miss Charleston completely."
   (b) What is $A \cap B$?
   (c) What is $A \cup B$?
   (d) What is A^c?

7. Suppose you know the distribution of the passing distances of icebergs from the anchor location of the research vessel in Exercise 1(d) and that you can evaluate

   $$P[\text{"Iceberg is within } x \text{ km of the vessel."}], \quad \text{for any } x \geq 0.$$

   (a) State whether $A \subset B$ for the following two event definitions
      A: "Iceberg is within 0.5 km of the vessel."
      B: "Iceberg is within 1.5 km of the vessel."
   (b) State how you would evaluate the probability of "getting a good scare but not having to move the vessel", i.e., of the event $C = B - A$. Identify the probability rules which permit you to evaluate $P[C]$ as you have said.

8. Suppose $f(s) = e^{-s}$, for $s \geq 0$, is the probability density function for the closest distance of a rainband to the location of an outdoor graduation ceremony.

   (a) Find the probability that the rainband comes within 3 km of the ceremony.

   (b) Find the probability that the rainband is not closer than 1 km.

   (c) Use (4.16) and your answer to (b) to find the probability that the rainband does come closer than 1 km.

9. Diagram the events E_2, E_3, E_4, and E_5, defined in Section 4.4.

10. Use data from the CLIMAT data file, for Lander, WY (LND) for July and August.

   (a) Plot average dewpoint temperature against average wind speed.

   (b) Create the basis for a bivariate histogram or contour plot by choosing equal interval demarcations on both axes and writing within each rectangle the number of points within its boundaries. Photocopy your product.

   (c) Overwrite a pale copy of (b) with the relative frequencies for each rectangular "interval", obtained by dividing each frequency by 62: the total number of days of data. Then hand contour this plot.

11. On a second copy of the product of Exercise 10(b), outline the events

$$A: ``\overline{TD} < 80" \qquad \text{and} \qquad B: ``\overline{W} > 8".$$

   (a) Find the relative frequencies of A and B: $\hat{P}[A]$ and $\hat{P}[B]$.

   (b) Find the relative frequency of $A \cup B$; compare with $P[A] + P[B]$; and account for the difference.

   (c) Use the results of (a) to find the relative frequencies of

$$A^c: ``\overline{TD} \geq 80" \qquad \text{and} \qquad B^c: ``\overline{W} \leq 8".$$

   (d) Is there evidence that these variables are statistically independent? (Hint: See (4.20) and compare $P[A] \times P[B]$ with $P[A \cap B]$.)

   (e) Find the conditional probability that the dewpoint temperature is less than 80, given that the average wind speed exceeds 8.

   (f) Suppose that on a particular day in summer you know that $\overline{W} > 8$. Use the result of (d) and Bayes' Theorem to assign probabilities to A and A^c, conditioned on this information.

5

STOCHASTIC VARIABLES AND THE IDENTIFICATION OF THEIR DISTRIBUTIONS
"distilling uncertainty"

5.1 INTRODUCTION

In the previous chapter variables were discussed in the contexts of probability models. In particular, our focus was on:

(i) the definition of a description space for the possible values of
each variable,
(ii) characterization of events whose probabilities can be assessed, in
terms of those variables.
(iii) properties of rules for assigning probabilities to events.

Here we get specific about what "stochastic variable" means and about
different ways of identifying a probability model. Then we apply these
concepts to enumeration or "counting" variables.

The question, "What is a variable?" may seem like a trivial question.
However, since "variable" has been used in connection with describing
possible states of a geophysical system as well as outcomes of studies of
system properties, there is some fuzziness in its use which needs to be
cleared before considering the probability distribution of a variable. A
full-scale description of a geophysical system, even at a single location and

time, includes information about many system parameters. This description is much more detailed than we can use in any scientific inference process. The information we will use in answering the scientific question we have posed for ourselves is generally a reduction of the details available to us. We will assume that it is either a vector which records a select few of the system parameters or a univariate summary, such as a daily average. This, rather than the many-faceted description, is our *variable*. When we can associate probabilities with the possible values of the variable, i.e., the vector or scalar values which may be observed, we call it a *random* or *stochastic variable*; and the rule for assigning probabilities to its values is called a *probability* or *stochastic distribution*. Here we make a transition from using s to denote a generic outcome or possible observation to using the notations X and x. The capital letter stands in place of the name of the variable, of which examples are temperature, salinity, maximum wave height, and number of successful cloud seedings. The lower-case letter records or indexes specific possible values, such as $-53°C$, 110 ppm, 3.8 m, or 10 of 15. We shall continue to use S to denote the collection of all possible values, especially for discrete variables.

We have given separate consideration to distributions of discrete-valued variables and distributions of continuous-valued variables. In the first case the distribution can be specified by the collection of probabilities corresponding to each possible value

$$\{p(x): x \in S\}, \quad \text{with } p(x) \geq 0 \quad \text{for all } x \quad \text{and} \quad \sum_{x \in S} p(x) = 1. \quad (5.1)$$

This collection—the rule which assigns probabilities—is called a *probability mass function* (p.m.f.). In the case of a continuous-valued variable the distribution may be specified by a *probability density function* (p.d.f.). A p.d.f. is a bounded, non-negative, continuous function $f(x)$ which we have defined for every observable value of the variable whose stochastic behavior it describes and which we use for evaluation of probabilities of the events of interest, as

$$P[E] = \int_E f(x) \, dx. \quad (5.2)$$

In both the discrete and continuous cases we noted that the description space S may be multidimensional, so that each element x is a vector. For the most part, the concepts introduced in this chapter and the next will be written in the relatively simple notation of a scalar variable. Their analogues for vector-valued variables will be found in the exercises.

Expressions (5.1) and (5.2) give us vehicles for assessing probabilities for possible outcomes to the observation of atmosphere or ocean phenomenon. The keys in these two expressions are the p.m.f. in (5.1) and the p.d.f. in (5.2), with which the Exercises of Chapter 4 have given you some experience. Here we present alternative characterizations for probability models. These are equivalent, in the sense that if we know what the p.m.f. or p.d.f. for a variable looks like, then we also know what its "characteristic function" (ch.fn.) looks like. This is true in the reverse, as well. That is, if we know the ch.fn. then we have full identification of the probability model, including the p.m.f. or p.d.f.

Why do we need more than one way of describing a probability model? The answer is that the characteristic function gives us handy ways of getting at properties of distributions and linking distributions of variables. To begin to explore these possibilities, think about the fact that a "probability distribution" describes the way in which a given physical situation apportions likelihoods among possible outcomes, in terms of the values of the variable we have assigned to register outcome. Thus *it distributes probability*. If the variable is a scalar, this distribution of probability is over the axis of real numbers. Think about it as the distribution of a unit of mass. Writing its probability mass function or probability density function is like drawing a picture of it. With this image in mind, we can see that the weighted average

$$\mu = \sum_{x \in S} x\, p(x) \quad \text{or} \quad \mu = \int_{-\infty}^{+\infty} x\, f(x)\, dx \quad (5.3)$$

is the balance point; and the weighted average squared increment

$$\sigma^2 = \sum_{x \in S} (x - \mu)^2 p(x) \quad \text{or} \quad \sigma^2 = \int_{-\infty}^{+\infty} (x - \mu)^2 f(x)\, dx \quad (5.4)$$

is the squared dispersion about this central value. We call μ in (5.3) the *mean of the distribution*, and σ^2 in (5.4) the *variance of the distribution*. These are the two most often noted features in the characterization of the distribution of a variable. Nonetheless they do not fully characterize it, as we can see by referring back to Fig. 3.6. To be complete we need to know more than just the mean and variance.

From calculus we know that any finite, bounded, summable, or integrable function is fully specified by all its finite integer moments

$$\mu_k = \sum_{x \in S} x^k p(x) \quad \text{or} \quad \mu_k = \int_{-\infty}^{+\infty} x^k f(x)\, dx. \quad (5.5)$$

So, specifying the complete collection of μ_k is another way of fully describing a distribution. This sounds messy and awkward. However, the "characteristic function" does exactly that in one neat formula. Observe that if we write

$$\Phi(\xi) = \sum_{x \in S} e^{i\xi x} p(x) \quad \text{or} \quad \Phi(\xi) = \int_{-\infty}^{+\infty} e^{i\xi x} f(x)\, dx \quad (5.6)$$

then the kth derivative of Φ evaluated at $t = 0$ is precisely

$$\frac{d^k}{d\xi^k}\Phi(\xi)\bigg|_{\xi=0} = \sum_{x \in S}(ix)^k e^{i\xi x}p(x)\bigg|_{\xi=0} = i^k\sum_{x \in S}x^k p(x) \qquad (5.7)$$

for a discrete distribution; and

$$\frac{d^k}{d\xi^k}\Phi(\xi)\bigg|_{\xi=0} = \int_{-\infty}^{+\infty}(ix)^k e^{i\xi x}f(x)\,dx\bigg|_{\xi=0} = i^k\int_{-\infty}^{+\infty}x^k f(x)\,dx \quad (5.8)$$

for a continuous distribution. Thus we have a conveyance for all the integer moments, in a single function $\Phi(\xi)$. And this we call the *characteristic function* for the distribution of the variable X. We have written it as a function of ξ. In any specific application, it will also be a function of whatever parameters assist in describing the distribution, as we shall soon see.

Because we will be linking distributions for compositions of variables, such as the descriptive statistics studied in Chapter 3, characteristic functions will be key tools. They obviate the necessity of doing integrals which would be absolutely daunting, even numerically; and they reveal the simplistic beauty of probability structures.

Before looking at specific distributions we introduce one final generic concept: the concept of "statistical expectation". For a stochastic variable, or a function of a stochastic variable, this denotes the average value where the averaging is with respect to the distribution of the variable. It is a weighted average, with the weights assigned by the distribution of probability over S or $(-\infty, +\infty)$. If X is the variable assigned to record an outcome, the *statistical expectation* of X itself is just μ, the mean of the distribution; and we write

$$\mu = \mathscr{E}[X].$$

If the function of the variable is the squared increment of X from the mean of its distribution, $(X - \mu)^2$, the *statistical expectation* is the variance; and we write

$$\sigma^2 = \mathscr{E}\left[(X - \mu)^2\right].$$

In general, for any function of the variable, say $h(X)$, whose product with the p.m.f. or p.d.f. is summable or integrable over the set of all possible values of the variable, the *statistical expectation* is defined as

$$\mathscr{E}[h(X)] = \sum_{x \in S}h(x)p(x) \qquad \text{or} \qquad \mathscr{E}[h(X)] = \int_{-\infty}^{+\infty}h(x)f(x)\,dx.$$

$$(5.9)$$

Because its argument is a stochastic variable, the values $h(X)$ takes on will be governed by the distribution that governs X: Thus it too is a stochastic variable. In the same sense that the mean μ describes a weighted average value of X, $\mathscr{E}[h(X)]$ describes a weighted average value of $h(X)$. Simple examples that will be of use to us in the following are $h(X) = X(X-1)$ with

$$\mathscr{E}[X(X-1)] = \sum_{x \in S} x(x-1)p(x)$$

or   (5.10)

$$\mathscr{E}[X(X-1)] = \int_{-\infty}^{+\infty} x(x-1)f(x)\,dx$$

and $h(X) = \exp(i\xi X)$ for which the statistical expectation is the characteristic function for the distribution of X:

$$\mathscr{E}[e^{i\xi X}] = \sum_{x \in S} e^{i\xi x}p(x) = \Phi(\xi)$$

or   (5.11)

$$\mathscr{E}[e^{i\xi X}] = \int_{-\infty}^{+\infty} e^{i\xi x}f(x)\,dx = \Phi(\xi).$$

5.2 THE BINOMIAL DISTRIBUTION

In studying the specific rules for assigning probabilities to outcomes, we begin with the distributions of enumeration variables, i.e., variables which record counts of the number of occurrences of a designated atmosphere or ocean phenomenon in a fixed period of time. Conditions under which a distribution is appropriately described by a Binomial probability mass function will be defined first. The classical designation for this important distribution is introduced in terms of its descriptive parameters; and some of its most useful properties are considered. This section will set a pattern for exploration of other distributions. (It will help you in thinking about any specific probability designation, to note and adopt this pattern.)

The Binomial probability distribution pertains to circumstances in which there are a fixed number of opportunities for the occurrence of an atmospheric or oceanic phenomenon, say, for example, the occurrence of significant precipitation or the presence of ice floes within 100 yards of a research vessel, in a given 24-hour period. For each opportunity, that is, for each 24-hour period of observation, the occurrence or non-occurrence is recorded. If there are N opportunities, the enumeration variable has possible values $0, 1, 2, \ldots, N$. *Provided that outcomes for distinct opportunities are independent of one another and provided the probability of the occurrence is the same at each opportunity* then the number of occurrences has a *Binomial distribution*. The common probability for each possible occurrence is denoted by p; and we have used N to denote the number of opportunities. Thus there are two constants associated with the distribution: p and N. We write $\mathscr{B}(N, p)$ to designate the Binomial distribution, and call N and p the parameters of this distribution.

In order to understand references to this distribution in the literature of your field of study and to apply it to different types of situations, it will help to learn the translation to the standard (classical) statistical terminology for a Binomial distribution. In these terms, an opportunity for the occurrence or non-occurrence of a designated phenomenon is referred to as "an experiment with just two possible outcomes". This is sometimes called a *Bernoulli trial*. Reflecting on its history in connection with gambling, one of the possible outcomes is designated "success" and the other as "failure". The probability associated with the success outcome is denoted by p. Because there are only two possible outcomes, the probability associated with non-success, or failure, is $(1 - p)$. If N Bernoulli trials are conducted, each one independent of the others in the sense that the outcome of trial k is in no way influenced by the outcomes of any other trials, and if the number of successes in the sequence of trials is recorded as

X = number of successes in N independent Bernoulli trials,

then the variable X has a Binomial distribution.[1] We write $X \to \mathscr{B}(N, p)$.

The probabilities this distribution assigns to values in the outcome space for X are

$$P[X = x] = \begin{cases} \dfrac{N!}{x!(N-x)!} p^x (1-p)^{N-x}, & \text{for } x = 0, 1, \ldots, N \\ 0, & \text{for all other real numbers.} \end{cases}$$

$$(5.12)$$

To see that this is derived from the defining conditions of a Binomial variable, consider that the probability multiplication rule for independent outcomes assigns probability $p^x(1-p)^{N-x}$ to each possible sequence of x successes and $(N-x)$ failures, regardless of the order in which they occur. The enumeration variable does not distinguish sequences with different orderings of successes and failures; it records only the number of successes. Accordingly, the collective probability of exactly x successes is the number of different orderings of x successes and $(N-x)$ failures, times their common probability. The number of different orderings is given by the *combinatorial coefficient*

$$\binom{N}{x} = \frac{N!}{x!(N-x)!}$$

---

[1] Convention and logical clarity recommend the use of a capital letter for the variable whose distribution we are discussing and a lower-case letter to index specific values in the outcome space.

which is defined for $x = 0, 1, 2, \ldots, N$, with $0! = 1$ and $x! = x(x-1)!$ for $x > 0$.[2] Thus the desired probability is the product

$$\binom{N}{x} p^x (1-p)^{N-x}.$$

Any real number which is not one of the integers 0 to N is not a possible outcome for X and consequently is assigned 0 probability. Together these establish (5.12) as the probability mass function for the Binomial distribution. We may confirm that the probabilities sum to one, by using the equivalence

$$\sum_{x=0}^{N} \binom{N}{x} a^x b^{N-x} = (a+b)^N.$$

Thus, with $a = p$ and $b = (1-p)$,

$$\sum_{x=0}^{N} \binom{N}{x} p^x (1-p)^{N-x} = [p + (1-p)]^N = 1.$$

To construct an illustration of a Binomially distributed variable, let's suppose we are involved in analyzing the results of a cloud seeding experiment which has been carefully planned and executed. The objective of seeding clouds is to produce rainfall with a significantly higher success rate than is expected if the clouds are not seeded. To control variability which would obstruct determination of seeding effects, only one specific type of cloud cover is designated as a candidate for seeding. Let's say that we know from past cloud precipitation records that the likelihood of naturally occurring rainfall from this type of cloud cover is $p = 0.24$. Thus if the seeding procedure does not change the likelihood of rainfall, that is, if it has no effect in enhancing precipitation, then each occasion of cloud seeding is a Bernoulli trial with probability 0.24 of success. N such occasions, on which the cloud cover was of the type specified in the experimental design and cloud seeding was done, produce a record of successes and failures. Provided that we are justified in making the assumption that the outcomes of successive seedings are independent of

---

[2]You may recall combinatorial coefficients from algebra where they are used to give the coefficients in the expansion of $(a+b)^N$. What you will want to remember for use here is that for small N, they can be easily found from the $(N+1)$st row of Pascal's triangle

$$
\begin{array}{ccccccc}
 & & & 1 & & & \\
 & & 1 & & 1 & & \\
 & & 1 & 2 & & 1 & \\
 & 1 & & 3 & 3 & & 1 \\
 & & & \vdots & & & \\
\end{array}
$$

Here each row begins and ends with 1 and intermediate values are the sum of the bracketing values of the preceding row.

one another, the number of times that seeding was followed by significant precipitation will be a Binomial variable with parameters N and p.

Comparing our assumptions with the defining assumptions of a Binomial distribution and looking back at (5.12), we see that specifying the number of trials N and the probability of success on each trial p is sufficient to give values to the probabilities of all $(N + 1)$ possible outcomes. When N is quite small, the individual probabilities can be easily computed longhand. For larger values, as we find in most practical applications, the values may be obtained with a statistical software package. For example, the MINITAB command "PDF; BINOMIAL N, p." returns all of the values of $P[X = x]$ which have at least four significant digits. We will return to this illustration below.

We next turn to establishing the key features of the generic Binomial distribution, namely, its mean, variance, and characteristic function. We will prove that they are as given in

ASSERTION 5.1.   *For the Binomial distribution $\mathscr{B}(N, p)$*

(a) *The mean is $\mu = Np$.*
(b) *The variance is $\sigma^2 = Np(1 - p)$.*
(c) *The characteristic function is $\Phi(\xi) = [pe^{i\xi} + (1 - p)]^N$.*

Proof.   Anticipating their use in establishing our claim, we recall that

$$\sum_{y=0}^{M} \binom{M}{y} a^y b^{M-y} = (a + b)^M$$

for all a and b, and integer $M > 0$; and note that for $x > 0$

$$x\binom{N}{x} = x\frac{N!}{x!(N-x)!} = x\frac{N(N-1)!}{x(x-1)!(N-x)!}$$

$$= N\frac{(N-1)!}{(x-1)![(N-1)-(x-1)]!} = N\binom{N-1}{x-1}$$

and for $x > 1$

$$x(x-1)\binom{N}{x} = x(x-1)\frac{N(N-1)(N-2)!}{x(x-1)(x-2)!(N-x)!}$$

$$= N(N-1)\frac{(N-2)!}{(x-2)![(N-2)-(x-2)]!}$$

$$= N(N-1)\binom{N-2}{x-2}.$$

With these we can easily establish the three elements of the claim.

(a)

$$\mu = \mathscr{E}[X] = \sum_{x=0}^{N} x \binom{N}{x} p^x (1-p)^{N-x}$$

$$= 0 + \sum_{x=1}^{N} N \binom{N-1}{x-1} p^1 p^{x-1} (1-p)^{(N-1)-(x-1)}$$

$$= Np \sum_{(x-1)=0}^{N-1} \binom{N-1}{x-1} p^{x-1} (1-p)^{(N-1)-(x-1)}.$$

Taking $M = N - 1$ and $y = x - 1$, we may write this as

$$\mu = Np \sum_{y=0}^{M} \binom{M}{y} p^y (1-p)^{M-y} = Np[p + (1-p)]^M = Np.$$

(b) To evaluate

$$\sigma^2 = \mathscr{E}\left[(x-\mu)^2\right] = \mathscr{E}[X^2] - \mu^2$$

we first write

$$\mathscr{E}[X^2] = \mathscr{E}[X(X-1) + X] = \mathscr{E}[X(X-1)] + \mathscr{E}(X)$$

and evaluate the term

$$\mathscr{E}[X(X-1)] = \sum_{x=0}^{N} x(x-1) \binom{N}{x} p^x (1-p)^{N-x}$$

$$= 0 + 0 + \sum_{x=2}^{N} N(N-1) \binom{N-2}{x-2} p^2 p^{x-2} (1-p)^{(N-2)-(x-2)}$$

$$= N(N-1) p^2 \sum_{(x-2)=0}^{N-2} \binom{N-2}{x-2} p^{x-2} (1-p)^{(N-2)-(x-2)}.$$

Taking $M = N - 2$ and $y = x - 2$, this is

$$\mathscr{E}[X(X-1)] = N(N-1) p^2 \sum_{y=0}^{M} \binom{M}{y} p^y (1-p)^{M-y} = N(N-1) p^2.$$

Since we already have $\mu = \mathscr{E}[X] = Np$, we put the pieces together to get

$$\sigma^2 = \mathscr{E}[X(X-1)] + \mathscr{E}(x) - \mu^2$$
$$= N(N-1) p^2 + Np - (Np)^2$$
$$= -Np^2 + Np$$
$$= Np(1-p).$$

(c) Evaluating the characteristic function is the easiest of the three:

$$\Phi(\xi) = \mathscr{E}[e^{i\xi x}] = \sum_{x=0}^{N} e^{i\xi x} \binom{N}{x} p^x (1-p)^{N-x}$$

$$= \sum_{x=0}^{N} \binom{N}{x} (pe^{i\xi})^x (1-p)^{N-x} = [pe^{i\xi} + (1-p)]^N. \quad\blacksquare$$

If we had established the characteristic function first, we might have obtained $\mathscr{E}[X]$ and $\mathscr{E}[X^2]$ directly from $\Phi(\xi)$, as

$$\mathscr{E}[X] = \frac{1}{i} \frac{d\Phi}{d\xi}\bigg|_{\xi=0} \quad \text{and} \quad \mathscr{E}[X^2] = \frac{1}{i^2} \frac{d^2\Phi(\xi)}{d\xi^2}\bigg|_{\xi=0}.$$

These will give the same parametric functions for μ and $\sigma^2 = \mathscr{E}[X^2] - \mu^2$, as you can easily confirm. (See Exercise 7(c).)

For the evaluation of conjectures about parameter values, for this and other distributions, we will wish to evaluate probabilities of the "tails of the outcome spaces" for specified values of p, say $p = p_0$. That is, we ask and answer the questions:

What is the probability of observing X as small as x_l, if p has the value p_0?

What is the probability of observing X as large as x_u, if p has the value p_0?

If p has value p_0, then we generally expect to observe X near the center of the distribution $\mathscr{B}(N, p)$, i.e., among the values with the largest probabilities. If we observe a value of X away from the center of this distribution, it suggests that we have mis-specified p: large observed values suggest that the true value of p is greater than p_0, and small values suggest that the true value of p is smaller than p_0. However, before we take scientific action on evidence that the conjecture $p = p_0$ may be invalid, we assess the probability of obtaining a value as far out in the tail of the distribution with the conjectured parameter value, as the value observed. Thus, we answer one or the other of the two questions above.

"As small as x_l" means "one of the values $\leq x_l$". So the answer to the first question is what we call the cumulative distribution function (c.d.f.) evaluated at x_l:

$$F(x_l) = \sum_{x=0}^{x_l} \binom{N}{x} p_0^x (1-p_0)^{N-x}.$$

"As large as x_u" means "one of the values $\geq x_u$". So the answer to the second question is

$$\sum_{x=x_u}^{N} \binom{N}{x} p_0^x (1-p_0)^{N-1} = 1 - \sum_{x=0}^{x_u-1} \binom{N}{x} p_0^x (1-p_0)^{N-x} = 1 - F(x_u - 1).$$

Again we may obtain the required numbers with the aid of MINITAB. The command "CDF; BINOMIAL N, p." returns the c.d.f. for any specified pair of parameter values.

Returning to the cloud seeding example, we will evaluate the expected value of the number of occasions on which there would be significant precipitation if seeding did not alter the probability of precipitation. Then we will find and discuss a "tail probability" for a hypothetical observation. Let's say that we have designed our research to include 12 seeding days with similar cloud cover, with the days sufficiently separated in time that the seeding on any one does not effect the outcome on another; and we wait until the results are all in, before assessing seeding effect. Our design meets the defining conditions of the Binomial distribution with parameters $N = 12$ and $p = 0.24$. Consequently, if seeding does not have an impact on the probability of precipitation, the expected value of the number of occurrences of significant precipitation following seeding, among the 12 opportunities, is

$$\mathscr{E}[X] = 12(0.24) = 2.88 \quad \text{or} \quad \text{approximately 3.}$$

Suppose now that, when the results are all in, the experimental outcome is actually $x = 6$. This is on the large side, if seeding has not increased the probability of precipitation from 0.24. However, it is possible, and we can evaluate the probability of actually having as many as this, if the precipitation probability is unaltered. Read "as many as 6" as "6 or more" and compute

$$P_r[X \geq 6 | p = 0.24] = \sum_{x=6}^{12} \binom{12}{x} (0.24)^x (0.76)^{12-x} = 1 - F(5).$$

Using MINITAB, we find this "tail probability" to be $1 - 0.955 = 0.045$.

We will focus on the derivation of scientific decisions from tail probabilities and assessments of relative likelihoods in the chapter on hypothesis testing. For the present we simply note that the probability of having as many as 6 occurrences of significant precipitation as an outcome, if p is really 0.24, is extremely small. It would be easier to support the hypothesis that seeding had increased the probability of precipitation under the cloud cover conditions of the experiment.

5.3 THE POISSON DISTRIBUTION

A major focus in defining the conditions under which a variable has a Binomial distribution was on the fixed, finite number of opportunities for the occurrences of the phenomenon being counted. In contrast, a Poisson distributed variable can be thought of as a count arising from infinitely many opportunities. Specifically, we can think of the full interval or region of observation as a continuum of opportunities for occurrences that can take place at any moment or point of time or space. Thus, a month is regarded as a continuous period in distinction to being a collection of 30 bits; and an ocean basin is a physically continuous region rather than a finite number of grid squares. Again we will let

Y = number of occurrences during the span of observation;

although, here, we note that there is no largest integer which cannot be exceeded by Y. The possible outcome values are written $0, 1, 2, \ldots$.

Standard statistical terminology generally defines the Poisson as the distribution of the number of "events" occurring during a predetermined period. Ice storms, tornados, or meteorite sightings are examples we might wish to consider as Poisson distributed events. To derive the distribution, we assume that the mean rate of occurrence is constant and relatively small. We also place a requirement of independence on non-overlapping subintervals or regions.

Formally, we say that *Y has a Poisson distribution provided that Y records the number of events which occur during interval t with constant mean rate of occurrence λ/unit time* **and** *all of the following hold.*

(a) The probability that one event occurs in any subinterval of length Δt is proportional to Δt, which we write as

$$p_1(\Delta t) = \lambda \Delta t + \mathcal{O}_1(\Delta t).^3$$

(b) The probability of two or more events occurring in any subinterval of length Δt is negligible compared to the probabilities of none or one, which we write as

$$\sum_{y=2}^{\infty} p_y(\Delta t) = \mathcal{O}_2(\Delta t).$$

(c) The number of events occurring within any interval is independent of the number of events occurring within any other non-overlapping interval.

---

[3] The notation $\mathcal{O}(\Delta t)$ represents terms in Δt whose ratio to Δt has limit 0, as $\Delta t \to 0$; i.e., $\text{limit}_{\Delta t \to 0}\,[\mathcal{O}(\Delta t)/\Delta t] = 0$. Clearly, the sum of any finite number of these $\sum_m \mathcal{O}_m(\Delta t)$ is yet another $\mathcal{O}(\Delta t)$; and signs may be subsumed in the notation.

For the Poisson distribution there is only the single, mean rate parameter to specify. $\mathscr{P}(\lambda)$ is the shorthand designation for this distribution and we write $Y \to \mathscr{P}(\lambda)$ to say that Y is a Poisson distributed variable with mean rate parameter λ. The probabilities assigned by the Poisson distribution to values in the outcome space for Y are given by

$$P[Y=y]=p_y(t) = \begin{cases} \dfrac{(\lambda t)^{y}}{y!}e^{-\lambda t}, & \text{for } y=0,1,2,\ldots \\ 0, & \text{for all other values of } y. \end{cases} \tag{5.13}$$

We will demonstrate that (5.13) follows from the defining conditions above, confirm that the positive terms of (5.13) sum to one, and derive the basic properties of the Poisson distribution. For these purposes we first find an expression for $p_0(\Delta t)$, using condition (a) together with the required

$$\sum_{y=0}^{\infty} p_y(\Delta t) = 1.$$

Thus

$$p_0(\Delta t) = 1 - p_1(\Delta t) - \sum_{y=2}^{\infty} p_y(\Delta t) = 1 - \lambda\Delta t - \mathscr{O}_1(\Delta t) - \mathscr{O}_2(\Delta t)$$

$$= 1 - \lambda\Delta t - \mathscr{O}_0(\Delta t).$$

For any interval length specified in the argument, we will use the notation $p_y(\ )$ to denote the probability that y events occur during whatever interval appears in the argument. Now, if we consider any interval of length $t + \Delta t$ as the union of disjoint subintervals of length t and Δt, so that condition (c) guarantees the independence of occurrences of events in these subintervals, we may write

$$\Pr[(j-k) \text{ events in } t \text{ and } k \text{ events in } \Delta t]$$

$$= \Pr[(j-k) \text{ events in } t] \times \Pr[k \text{ events in } \Delta t]$$

$$= p_{j-k}(t)p_k(\Delta t).$$

The factorization is valid for any t and $\Delta t > 0$, $j \geq 0$, and $k \leq j$. Now, since j events in $t + \Delta t$ can occur in exactly one of the following mutually exclusive ways,

$$j \text{ in } t \quad \text{and} \quad 0 \text{ in } \Delta t$$

$$j-1 \text{ in } t \quad \text{and} \quad 1 \text{ in } \Delta t$$

$$\vdots$$

$$j-k \text{ in } t \quad \text{and} \quad k \text{ in } \Delta t$$

$$\vdots$$

then we can write its probability as the summation

$$p_j(t + \Delta t) = \sum_{k=0}^{j} \Pr[(j - k) \text{ events in } t \text{ and } k \text{ events in } \Delta t]$$

$$= \sum_{k=0}^{j} p_{j-k}(t) p_k(\Delta t)$$

$$= p_j(t) p_0(\Delta t) + p_{j-1}(t) p_1(\Delta t) + \sum_{k=2}^{j} p_{j-k}(t) p_k(\Delta t)$$

$$= p_j(t) [1 - \lambda \Delta t + \mathcal{O}_0(\Delta t)]$$

$$+ p_{j-1}(t) [\lambda \Delta t + \mathcal{O}_1(\Delta t)] + \mathcal{O}_2(\Delta t).$$

Dividing through by Δt, we get

$$[p_j(t + \Delta t) - p_j(t)] / \Delta t = -\lambda p_j(t) + \lambda p_{j-1}(t)$$

$$+ [\mathcal{O}_0(\Delta t) + \mathcal{O}_1(\Delta t) + \mathcal{O}_2(\Delta t)] / \Delta t.$$

Thus, provided that j is at least 1,

$$\frac{dp_j(t)}{dt} = \lim_{\Delta t \to 0} [p_j(t + \Delta t) - p_j(t)] / \Delta t = -\lambda p_j(t) + \lambda p_{j-1}(t). \quad (5.14)$$

For $j = 0$,

$$p_0(t + \Delta t) = p_0(t) p_0(\Delta t) = p_0(t) [1 - \lambda \Delta t + \mathcal{O}_0(\Delta t)]$$

and

$$[p_0(t + \Delta t) - p_0(t)] / \Delta t = -\lambda p_0(t) + \mathcal{O}_0(\Delta t) / \Delta t$$

for

$$\frac{dp_0(t)}{dt} = -\lambda p_0(t).$$

This last differential equation is easily seen to be satisfied by

$$p_0(t) = e^{-\lambda t}.$$

Now, taking $j = 1$ in (5.14) and substituting for $p_0(t)$, we get

$$\frac{dp_1(t)}{dt} = -\lambda p_1(t) + \lambda e^{-\lambda t}$$

which is easily shown to have solution

$$p_1(t) = \frac{\lambda t}{1!} e^{-\lambda t}$$

by direct differentiation.

To confirm the expressions for $j > 1$, we use the induction hypothesis

$$p_{j-1}(t) = \frac{(\lambda t)^{j-1}}{(j-1)!}e^{-\lambda t}$$

which we know to be true for $(j-1) = 1$, and prove that

$$p_j(t) = \frac{(\lambda t)^j}{j!}e^{-\lambda t}$$

satisfies (5.14):

$$\frac{d}{dt}\left[\frac{(\lambda t)^j}{j!}e^{-\lambda t}\right] = -\lambda\frac{(\lambda t)^j}{j!}e^{-\lambda t} + \lambda j\frac{(\lambda t)^{j-1}}{j!}e^{-\lambda t}$$

$$= -\lambda\left[\frac{(\lambda t)^j}{j!}e^{-\lambda t}\right] + \lambda\left[\frac{(\lambda t)^{j-1}}{(j-1)!}e^{-\lambda t}\right]$$

$$= -\lambda p_j(t) + \lambda p_{j-1}(t).$$

Thus, by induction, we have established (5.13) for all positive integers.

We can confirm that the positive terms sum to 1 by using the Taylor series expansion

$$e^z = \sum_{k=0}^{\infty}\frac{z^k}{k!}.$$

With $z = \lambda t$ and $k = y$,

$$\sum_{y=0}^{\infty}\frac{(\lambda t)^y}{y!}e^{-\lambda t} = e^{-\lambda t}\left[\sum_{y=0}^{\infty}\frac{(\lambda t)^y}{y!}\right] = e^{-\lambda t}e^{\lambda t} = e^0 = 1.$$

As an example of a Poisson distributed variable, imagine that you are part of an ocean research team, on a vessel anchored in the North Atlantic. The passage of an ice floe within 100 yards of the vessel is of great research potential for some members of the team. However, it is also a matter that requires evasive action if it appears to be headed for collision with the vessel. In the planning stages of the expedition you were told that you could expect an average of one floe every other day, coming within the 100-yard surveillance radius. Now you are out there on location, for a week, actually counting them. You will report the number of floes whose paths pass within 100 yards of the ship during this week. (If it becomes necessary to move the ship, we will assume that it will not be moved far enough to alter the parameters of the distribution of the enumeration variable.) With a hypothetical average of one every two days, the mean rate of floe intrusion, per week, is 3.5. Consequently, the number of these events that actually occur during the week is a Poisson variable with parameter $\lambda = 3.5$.

The Poisson distribution has just one parameter: the constant mean rate of occurrence of the events being counted. Of course, to evaluate probabilities of specific possible outcomes, it is essential to convert the mean rate to the mean number for the length of the time interval or the size of the region of observation: namely, to λt or λs. Having done that, the probabilities of all possible outcomes are easily obtained. Again, MINITAB is a handy source. The command "PDF; POISSON MEAN." using your value for MEAN: either λt or λs, will return all p.m.f. values greater than .0005.

We now derive the three basic properties, the mean, variance, and characteristic function, which describe the classic features of the distribution of the Poisson enumeration variable.

ASSERTION 5.2. *For the Poisson distribution of Y = number of occurrences of the enumerated event, in interval of length t, where λ is the mean rate per unit time*:

(a) The expected value of Y is $\mu = \mathcal{E}[Y] = \lambda t$.
(b) The expected value of Y^2 is $\mathcal{E}[Y^2] = (\lambda t)^2 + \lambda t$, for a variance of $\sigma^2 = \lambda t$.
(c) The characteristic function is $\Phi(\xi) = \mathcal{E}[e^{i\xi Y}] = e^{\lambda t(e^{i\xi}-1)}$.

Proof. The three properties are obtained quite simply, using the factorizations

$$y! = y(y-1)! \text{ for } y \geq 1, \quad \text{and} \quad y! = y(y-1)(y-2)! \text{ for } y \geq 2,$$

and the Taylor series expansion of e^z given previously.

(a)

$$\mu = \mathcal{E}[Y] = \sum_{y=0}^{\infty} y \left[\frac{(\lambda t)^y}{y!} e^{-\lambda t} \right]$$

$$= e^{-\lambda t} \left[0 + \sum_{y=1}^{\infty} y \frac{(\lambda t)(\lambda t)^{y-1}}{y(y-1)!} \right]$$

$$= e^{-\lambda t}(\lambda t) \sum_{(y-1)=0}^{\infty} \frac{(\lambda t)^{y-1}}{(y-1)!}$$

$$= e^{-\lambda t}(\lambda t)e^{\lambda t} = \lambda t.$$

(b) We write

$$\mathcal{E}[Y^2] = \mathcal{E}[Y(Y-1)] + \mathcal{E}[Y]$$

and find

$$\mathscr{E}[Y(Y-1)] = \sum_{y=0}^{\infty} y(y-1)\left[\frac{(\lambda t)^{y}}{y!}e^{-\lambda t}\right]$$

$$= e^{-\lambda t}\left[0+0+\sum_{y=2}^{\infty} y(y-1)\frac{(\lambda t)^{2}(\lambda t)^{y-2}}{y(y-1)(y-2)!}\right]$$

$$= e^{-\lambda t}(\lambda t)^{2}\sum_{(y-2)=0}^{\infty}\frac{(\lambda t)^{y-2}}{(y-2)!}$$

$$= e^{-\lambda t}(\lambda t)^{2}e^{\lambda t} = (\lambda t)^{2}.$$

Now, assembling the parts, we get

$$\mathscr{E}[Y^{2}] = (\lambda t)^{2} + \lambda t$$

as claimed, and finally

$$\sigma^{2} = \mathscr{E}[Y^{2}] - \mu^{2} = (\lambda t)^{2} + \lambda t - (\lambda t)^{2} = \lambda t.$$

(c)
$$\Phi(\xi) = \mathscr{E}[e^{i\xi Y}] = \sum_{y=0}^{\infty} e^{i\xi y}\left[\frac{(\lambda t)^{y}}{y!}e^{-\lambda t}\right]$$

$$= e^{-\lambda t}\sum_{y=0}^{\infty}\frac{(\lambda t e^{i\xi})^{y}}{y!}$$

$$= e^{-\lambda t}e^{\lambda t e^{i\xi}}$$

$$= e^{\lambda t(e^{i\xi}-1)}. \qquad \blacksquare$$

We may easily confirm that the first two derivatives of the characteristic function give us $\mathscr{E}[X]$ and $\mathscr{E}[X^{2}]$:

$$\frac{d}{d\xi}\Phi(\xi) = (i\lambda t e^{i\xi})e^{\lambda t(e^{i\xi}-1)}$$

so that

$$\frac{1}{i}\frac{d}{d\xi}\Phi(\xi)\Big|_{\xi=0} = \lambda t$$

and

$$\frac{d^{2}}{d\xi^{2}}\Phi(\xi) = \left[(i\lambda t e^{i\xi})^{2} + i^{2}\lambda t e^{i\xi}\right]e^{\lambda t(e^{i\xi}-1)}$$

so that

$$\frac{1}{i^2} \frac{d^2}{d\xi^2} \Phi(\xi) \Big|_{\xi=0} = (\lambda t)^2 + (\lambda t).$$

Returning to the example, imagine that during preparations for the research cruise you encountered skeptics on both sides of the question of the frequency of ice floe intrusions, although you regarded your best source of information to be the one that estimated an average of one every other day. Questions you might now like to answer are:

Q1. What is the probability that you count no more than one floe within a 100-yard radius of the ship, during the week?
Q2. What is the probability that you count at least 6?

The first is answered by evaluating the c.d.f. at $y = 1$; and, provided we believe $(\lambda t) = 3.5$, we can do this with the MINITAB command "CDF; POISSON 3.5." Since this command returns all the values shown in Table 5.1, not only do we have the answer to Question 1:

$$\Pr[Y \le 1 | t = 3.5] = 0.136,$$

we also have the information we need to answer Question 2.

For the latter, we obtain

$$\Pr[Y \ge 6] = 1 - \Pr[Y \le 5] = 1 - F(5) = 1 - 0.858 = 0.142.$$

If, in fact, you observe as few as 1 or as many as 6 during the week's period of observation, you may wish to question the reliability of your source and revise the guidance for future expeditions in this area. However, before taking scientific action, we will want to know more about

TABLE 5.1   Poisson Probabilities When the Mean Is 3.5

K	$P(X = K)$	$P(X \le K)$
0	0.0302	0.0302
1	0.1057	0.1359
2	0.1850	0.3208
3	0.2158	0.5366
4	0.1888	0.7254
5	0.1322	0.8576
6	0.0771	0.9347
7	0.0385	0.9733
8	0.0169	0.9901
9	0.0066	0.9967
10	0.0023	0.9990
11	0.0007	0.9997
12	0.0002	0.9999

assessment of relative likelihoods of possible observed values. In addition, it would be wise to bring in additional data from similar cruises, to use in conjunction with the single observation obtained here. Distributions for combinations of several independent Poisson variables will be considered in a later chapter.

EXERCISES

1. There are two sets of Arctic ozone measurements: one for the period 1935–1969 and another for the period 1984–1992. These daily measurements provide an excellent basis for comparison of the two periods.
   (a) Allowing for the strong annual cycle in ozone concentration, which is made evident by plotting monthly average values for successive months, identify the vector or scalar variable you would use to evaluate the change in ozone concentration from the earlier to the later time period.
   (b) If the two ozone measurement data sets were available in MINITAB file format, write the commands you would use to determine the value (vector or scalar) of the variable of part (a).

2. Refer to the scientific question in Exercise 2 of Chapter 1.
   (a) Describe the variable you would use to obtain an answer to this question.
   (b) Explain how this is a reduction of the information you would expect to be available to you.
   (c) Justify the use of this variable as the most appropriate summary of relevant information, given that you are constrained to base your inference on one (vector or scalar) variable.

3. Suppose that you have available the original salinity measurement records for Oceanographic Station #27, containing 45 years of daily values for 10 depths, from 0 to 175 meters.
   (a) With the objective of characterizing the difference between 50- and 150-meter salinities, including the annual cycle in this difference, identify the variable you will use for this job.
   (b) Use the salinity subfile of the STN#27 data set to evaluate the 45-year mean value for this variable.
   (c) What information do you need to evaluate the standard deviations of the elements of this variable? Is that information available from the STN#27 data set?

4. Use the distribution you constructed in Exercise 2 of Chapter 4 to evaluate the mean and variance for Variable = "number of days of significant snowfall in January". Do this by setting the possible values $1, 2, \ldots, 31$ into one column, the probabilities you calculated into

another column, and applying simple MINITAB arithmetic commands. In preparing your homework paper, declare an outfile that can be printed to show the commands you have given and the results of the MINITAB calculations.

Hint: If you have $x\,p(x)$ values in one column, say these are in C15, and $(x - \mu)^2\,p(x)$ values in another column, say C16, you can get the mean and variance as single number statistics with the commands

> LET K1 = MEAN(C15)
> LET K2 = SUM(C16)/(N(C16) − 1)
> PRINT K1, K2.

5. Consider a very simple variable, which we will call "the rain variable". This has value 1 if there is significant rain on the day in question and value 0 if there is not. (For this you need to take the attitude of the California farmer whose water economy is critical to his financial success; and forget about what may happen to the field hockey game you have planned for that day.) If the cloud conditions are such that our cloud almanac says the probability of rain is 0.24, evaluate the mean, variance, and characteristic function for the rain variable.

6. From CLIMAT for January, for each of the 15 stations, estimate p the probability of significant precipitation on any given day.

   Hint: p = number of days of significant precipitation in the month$/31$. Then set up a table in which the rows are labeled with the locations of the 15 stations and the columns with $p, \mu, \sigma, \Phi(\xi)$. Fill in the table, using estimates of p which you obtain from the data file.

7. Using the data file CLIMAT obtain estimates of the daily precipitation probabilities for both months January and July, separately for each of the five stations: ADQ, MFR, ABQ, CHS, CAR.

   (a) Plot these as number pairs $(p_{\text{JAN}}, p_{\text{JUL}})$ on a sketch of North America.

   (b) For CAR give the estimates of the mean, variance, and characteristic function for the number of days of significant precipitation in July.

   (c) Using the characteristic function,

   $$\frac{1}{i^k}\left[\frac{d^k}{d\xi^k}\Phi(\xi)\right]\Bigg|_{\xi=0} = \mathcal{E}[X^k]$$

   $$\mu = \mathcal{E}[X] \quad \text{and} \quad \sigma^2 = \mathcal{E}[X^2] - \mu^2,$$

   show that

   $$\mu = Np_{\text{JUL}} \quad \text{and} \quad \sigma^2 = Np_{\text{JUL}}(1 - p_{\text{JUL}}).$$

Evaluate these for Caribou, ME, with the value of p_{JUL} for CAR from part (a) substituted in the right-hand sides.

8. For each of the five locations ADQ, MFR, ABQ, CHS, CAR, estimate the daily probability of precipitation for April.
   (a) Use the MINITAB command "CDF;BINOMIAL N, P." with $N = 30$ and the estimated value of p, to get comparable precipitation distributions for the five locations.
   (b) Use scissors, tape, and a photocopier to present them on a single sheet.
   (c) Would you feel confident using the Binomial distribution to describe the stochastic behavior of the number of days of significant precipitation in April? If not, why not?
   Hint: Review the assumptions of the Binomial distribution.

9. Use the foregoing cloud seeding illustration of the use of the Binomial distribution.
   (a) If seeding is effective and increases the probability of precipitation to 0.52, what are the mean, variance, and characteristic function for the number of seedings out of 12 that are followed by significant precipitation?
   (b) Evaluate the probability of getting significant precipitation six or more of the times seeding is done under the circumstances described.

10. During the years 1953–1990, Florida had the highest tornado frequency, in number per year per 10^4 square miles, of any of the states of the U.S., with 8.3. Suppose we designate a 100×100 square mile area in central Florida, and assume that this 38-year statewide average frequency is characteristic of this area for the current year.
    (a) Write the defining assumptions of the Poisson distribution in terms of the variable $Y =$ "number of tornados in the designated area during the next 12 months" and state whether or not you think they are reasonable assumptions. If you do not believe they are reasonable, state your reasoning.
    (b) If the Poisson distribution is used to describe the stochastic behavior of Y, find the (i) mean value, (ii) variance and standard deviation, (iii) characteristic function for Y.
    (c) If we propose to enumerate the number of tornados in this surveillance area in central Florida for two consecutive years, what could we reasonably use to describe the distribution of the total number occurring? Are there any additional assumptions we would need to make? What will be the mean, variance, and characteristic function of this variable?

11. In an intense electrical storm, centered over a section of a national forest of uniform electrical attraction, the number of lightning strikes may average 3 per minute. Let the enumeration variable be the number of strikes within this section during a 5-minute period.

   (a) Write the defining assumptions of the Poisson distribution in terms of this variable and state whether or not you think they are reasonable assumptions.

   (b) Using the Poisson distribution to describe its stochastic behavior, give the (i) mean value, (ii) variance and standard deviation, (iii) characteristic function for the 5-minute enumeration variable.

   c) Now consider increasing the surveillance area to four times the size of the initial section. Identify the distribution of the (variable) number of lightning strikes within the larger area in a 1-minute period; and use MINITAB to generate the distribution. Label your printed OUTFILE clearly, identifying the information it contains, and attach it to your assignment papers.

6

THE EXPONENTIAL AND UNIFORM DISTRIBUTIONS
"describing uncertainty in time and space"

6.1 INTRODUCTION

The topics of this chapter logically follow our study of the distributions of variables which count events that occur at random in time or in space. What we mean by "at random in time or in space" is that we cannot predict with certainty when (or if) they will occur or at which specific locations, although we can assign a likelihood to each possibility. Examples we have used for illustration are occurrences of significant precipitation within a designated period of time and passages of ice floes within designated regions of the ocean. In this chapter we turn our attention to the related distributions of the waiting times and intervals between their occurrences.

Here we are focusing on the relative temporal/spatial structure of a system, as you would want to do if you were concerned about the probable duration of a drought, the proximity of a menacing phenomenon (such as the passage of an iceberg by a moored drilling rig), or the interval between destructive storms. We take the fact that these events occur as given and define our variable of interest to be the waiting time to the first occurrence, or the distance of its occurrence from a key location, or the interval between consecutive occurrences.

6.2 EXPONENTIAL DISTRIBUTIONS OF WAITING TIMES

We begin with the Exponential distribution, whose derivation is simple and delightfully direct. This is the distribution of the time between occurrences of Poisson distributed events. Accordingly, the assumptions we make regarding the circumstances of occurrences of the events with which we are concerned are precisely those from which we developed the Poisson distribution of the variable Y, recording the number of events during an interval of length t. Recall that we assume the mean rate of occurrence per unit time to be a constant, independent of t, which we denote by λ and that all of the following hold:

(a) The probability that one event occurs in any subinterval of length Δt is proportional to Δt; which we write as

$$p_1(\Delta t) = \lambda \Delta t + \mathscr{O}_1(\Delta t).$$

(b) The probability of two or more events occurring in any subinterval of length Δt is negligible compared to the probabilities of none or one, which we write as

$$\sum_{y=2}^{\infty} p_y(\Delta t) = \mathscr{O}_2(\Delta t).$$

(c) The number of events occurring within any interval is independent of the number of events occurring within any other non-overlapping interval.

Our treatment of the variable Y, in Chapter 5, regarded the interval or region of surveillance as fixed and focused on the number of events occurring within it. Here we have a different focus: We take the variable of interest to be *the interval between successive events* or *the waiting time to the next occurrence*.[1] We will call the variable T and index its possible values with t. The distribution of this interval variable may be derived from the distribution of Y. In this derivation we call attention to the length of the surveillance interval by writing the counting variable as Y_t, with a subscript which was implicit in Chapter 5. Thus Y_t denotes the number of occurrences within any designated interval of length t.

The derivation of the distribution of the interval variable T follows from the fact that there is only one way T can exceed any positive value t,

---

[1]An essential, but initially curious, feature of the distribution of T is that we may define T as starting from any point in time, with the same result. The timer on the interval which we shall record, to the next occurrence of the phenomenon of interest, can be started at the moment of the previous occurrence or it can be started whenever it is convenient to do so, within the research schedule. We will return to this topic and provide a rigorous argument, following the proof of Assertion 6.1. For the present it is noted in explanation of the apparent vagueness of the definition.

namely, for there to be zero occurrences of the enumerated events in the interval $(0, t]$. Thus

$$\Pr[T > t] = \Pr[Y_t = 0] = \frac{(\lambda t)^y}{y!} e^{-\lambda t} \Big|_{y=0} = e^{-\lambda t}.$$

We note now that the probability on the far left of this expression is just one minus the cumulative distribution function for T. Consequently,

$$F_T(t) = \Pr[T \le t] = 1 - \Pr[T > t] = 1 - e^{-\lambda t}$$

provided that $t > 0$. Since T denotes an interval in space or time, the probability that it is less than any number which is less than or equal to zero, must be zero. Thus we can write the c.d.f. for all real arguments as

$$F_T(t) = \begin{cases} 0, & \text{if } t \le 0 \\ 1 - e^{-\lambda t}, & \text{for } t > 0. \end{cases} \tag{6.1}$$

Because this is a continuous distribution, the probability density function gives us an alternative designation of its structure, namely

$$f_T(t) = \frac{dF_T(t)}{dt} = \begin{cases} 0, & \text{if } t \le 0,^2 \\ \lambda e^{-\lambda t}, & \text{for } t > 0. \end{cases} \tag{6.2}$$

Equations (6.1) and (6.2) are equivalent designations of the distribution we call *the Exponential distribution with parameter* λ. Since there is no finite value which cannot be exceeded by T, with at least some small probability, we denote the set of possible outcomes for the waiting time variable as $(0, \infty)$.

Summertime tornados reported for Kansas provide a good example of Poisson distributed events. The state has an average of 6 tornado reports during each 3-month, or 92-day, summer. Let's say that you are planning to visit the state as a tornado observer, with a base of operations in Rooks County. Undoubtedly, one primary concern in preparing for this project is the length of time from your arrival to the first tornado reported somewhere in the state. We can assess the probability that your wait will not exceed any given time interval by evaluating the cumulative distribution function for the length of that interval. In fact, with MINITAB it is possible to generate the c.d.f. for any collection of possible waiting times. Say we select integer numbers of days. We can either repeat the command "CDF t; EXPONENTIAL b.", where b is the reciprocal of the mean rate λ, for $t = 1, 2, 3, \ldots, 30$; or we can create a column in MINITAB that has these 30 integers in it, and give the CDF command once. The command

---

[2] Writing the density function with its value specified for all real arguments should not be regarded as a sign of unbridled compulsiveness. It will save considerable trouble in the future, in situations in which it is easy to forget that it is positive for only a portion of the real numbers.

sequence for the latter, with $b = (6/92)^{-1} = 15.3$, is

>SET C1.
>DATA 1 2 3 4 5 6 7 8 9 … 30
>DATA END.
>CDF C1; EXPONENTIAL 15.3.

to which MINITAB responds with

1	0.0633	11	0.5127	21	0.7465
2	0.1225	12	0.5436	22	0.7626
3	0.1781	13	0.5724	23	0.7776
4	0.2301	14	0.5995	24	0.7917
5	0.2788	15	0.6248	25	0.8049
6	0.3244	16	0.6486	26	0.8172
7	0.3671	17	0.6708	27	0.8288
8	0.4072	18	0.6916	28	0.8396
9	0.4447	19	0.7111	29	0.8497
10	0.4798	20	0.7294	30	0.8593

Thus for each full 24-hour day, through day 30, we have the probability that the waiting time for the first occurrence of a tornado will not exceed that number.

In addition to providing values of the c.d.f. for input values of t, MINITAB enables us to obtain inverse values. That is, suppose that you want to know what value of t would be exceeded with probability 0.25. We can write

$$0.25 = \Pr[T > t] = 1 - \Pr[T \le t] = 1 - F_T(t),$$

for which

$$F_T(t) = \Pr[T \le t] = 0.75,$$

and use the command "INVCDF 0.75; EXPONENTIAL 15.3." which returns "21.21". That is, the probability is 0.75 that your waiting time will not exceed 21.21 days.

We now establish the three basic descriptive properties for the Exponential distribution. Just as with the Poisson distribution from which it was derived, the distribution has a single parameter: the mean number of occurrences per unit time. Accordingly, all of its properties will also be described in terms of this one number.

ASSERTION 6.1.   *For the Exponential distribution with parameter λ, the mean, variance, and characteristic function are given by:*

(a) $\mu = \mathscr{E}[T] = 1/\lambda$
(b) $\sigma^2 = \mathscr{E}[(T - \mu)^2] = 1/\lambda^2$
(c) $\Phi(\xi) = \lambda(\lambda - i\xi)^{-1}$.

Proof. The proof follows easily from the fact that for a continuous distribution of a scalar variable, the expected value of any function of the random variable is obtained by integrating the product of this function and the p.d.f. of the variable over the whole real line:

$$\mathscr{E}[g(T)] = \int_{-\infty}^{+\infty} g(t) f_T(t)\, dt.$$

We will prove part (c) first and obtain (a) and (b) from the first two derivatives of the characteristic function.

$$\Phi_T(\xi) = \mathscr{E}[e^{i\xi T}] = \int_{-\infty}^{+\infty} e^{i\xi t} \left\{ \begin{matrix} 0, & \text{when } t \le 0 \\ \lambda e^{-\lambda t}, & \text{when } t > 0 \end{matrix} \right\} dt$$

$$= \int_0^\infty e^{i\xi t} (\lambda e^{-\lambda t})\, dt = \lambda \int_0^\infty e^{-t(\lambda - i\xi)}\, dt$$

$$= \lambda(\lambda - i\xi)^{-1}.$$

Now

$$\mathscr{E}[T] = \frac{1}{i} \frac{d\Phi_T(\xi)}{d\xi} \bigg|_{\xi=0} = \frac{1}{i} \lambda \left[i(\lambda - i\xi)^{-2} \right] \bigg|_{\xi=0} = \lambda/\lambda^2 = 1/\lambda$$

and

$$\mathscr{E}[T^2] = \frac{1}{i^2} \frac{d^2\Phi_T(\xi)}{d\xi^2} \bigg|_{\xi=0} = \frac{1}{i^2} \lambda \left[2i^2(\lambda - i\xi)^{-3} \right] \bigg|_{\xi=0} = 2\lambda/\lambda^3 = 2/\lambda^2$$

give us

$$\mu = 1/\lambda \quad \text{and} \quad \sigma^2 = \mathscr{E}[T^2] - \mu^2 = 2/\lambda^2 - (1/\lambda)^2 = 1/\lambda^2.$$

Confirmation of (a) and (b) is provided by Exercise 6.3. ■

We return to the claim made earlier that *the distribution of the waiting time to the next occurrence of the phenomenon being enumerated is independent of when the timer is started.* Here we give a rigorous proof of this initially surprising result and note some of its implications.

Proof of Claim: (i) Suppose that you begin observing a system without knowing the time of the last occurrence of the phenomenon. The probability that you wait at least time t_1 for the next occurrence is just the probability that it *does not* occur within the next interval of length t_1; i.e.,

$$\Pr[T > t_1] = e^{-\lambda t_1}, \qquad \text{for all } t_1 > 0.$$

(ii) Now suppose that you know that prior to the beginning of your observation period, time t_0 had elapsed since the last occurrence of the enumerated phenomenon. What you want now is the probability that you will wait an additional time t_1 for the next occurrence, conditioned on your prior knowledge. By the rule for conditional probabilities (4.21), this is

$$\Pr[T > t_0 + t_1 | T > t_0] = \Pr[T > t_0 + t_1] / \Pr[T > t_0]$$

$$= \int_{t_0 + t_1}^{\infty} \lambda e^{-\lambda t}\, dt \bigg/ \int_{t_0}^{\infty} \lambda e^{-\lambda t}\, dt$$

$$= e^{-\lambda(t_0 + t_1)} / e^{-\lambda t_0} = e^{-\lambda t_1}, \qquad \text{for all } t_0, t_1 > 0.$$

Since (i) and (ii) are identical, we have established the truth of the claim.

∎

You may feel more comfortable with this result when you recall condition (c) of the assumptions from which we derived the distribution of the interval variable. Specifically, the number of occurrences of the phenomenon of interest, in any interval, is independent of the number of its occurrences, in any non-overlapping interval. Here the intervals in question are $(0, t_0]$ and $(t_0 + t_1]$, and the numbers of occurrences postulated for each are both zero. The result says, simply, that "0 occurrences in $(0, t_0]$" has no influence on the probability of "0 occurrences in $(t_0 + t_1]$".

VIP: If the assumption of the independence of consecutive intervals is not a credible assumption for the system being observed, then the Exponential is not the right distribution for the interval variable T; and we need to return to the drawing board.

For the example of summertime tornados in Kansas, where the mean rate of occurrence is $\lambda = 6/92$ per day, the Exponential distribution gives us the intuitively correct expected waiting time for the next tornado since

$$\mathscr{E}[T] = 1/\lambda = 15.3.$$

This applies to the initial observing period, from arrival to the first tornado occurrence; and it applies to the waiting times between successive tornados, as well. If we define

$$T_j = \text{the interval between the } (j-1)\text{st and the } j\text{th tornados}$$

we have

$$\mathscr{E}[T_j] = 15.3, \qquad \text{for } j = 1, 2, 3, \dots.$$

Suppose your project requires data to be collected from 5 tornados. The statistical mean of the total interval of time you will need to spend in Kansas is

$$\mathscr{E}[T_1 + T_2 + T_3 + T_4 + T_5] = \sum_{j=1}^{5} \mathscr{E}[T_j] = 5(15.3) = 76.6 \text{ days.}$$

Note that this is the central value of the distribution of $\Sigma_j T_j$. It does not guarantee that you can actually complete the project in this period. Since the successive intervals are independent of one another, we know from Chapter 4 that the variance of $\Sigma_j T_j$ is $\Sigma_j \sigma_j^2$. Here this is

$$5\sigma_T^2 = 5(1/\lambda)^2 = 5(15.3)^2.$$

So the standard deviation of the total time you will need to complete the project is 34.3, which is more than a month on either side of the mean of 76.6 days!

6.3 UNIFORM DISTRIBUTIONS OF WAITING TIMES

Uniform distributions are the distributions of waiting times for events that are known to occur at fixed, regular intervals, but for which you do not know the point at which you commence being an observer *relative to* the interval between occurrences. Again the waiting time variable is a continuous variable, which we denote by T. However, here its value can never exceed the fixed length of the interval between occurrences. If we call this interval $(0, L]$, the waiting time for the next occurrence may be anywhere between L, if your arrival as an observer immediately follows the last occurrence of the event, to 0, if your arrival immediately precedes the next occurrence.

The derivation of the distribution of "the uniform waiting time variable" is simple and straightforward, as we show here. The two assumptions on which the distribution rests are:

(a) The events occur at fixed intervals of length L.
(b) We do not know the schedule for these events.

With these postulates we know that *it is just as likely that the point at which you arrive*, relative to their occurrences, *is within any subinterval of length* Δt, *as within any other of the same length* Δt. Thus, the probability of arriving within any interval of length Δt must be proportional to the interval length. We will let γ denote the constant of proportionality. Now

$$\Pr[t < T \le t + \Delta t] = \lambda \Delta t, \qquad \text{for all } t, t + \Delta t \in (0, L]$$

we may rewrite in terms of the cumulative distribution function of the waiting time variable, F_T, as

$$\lambda \Delta t = \int_t^{t+\Delta t} f_T(t)\, dt = F_T(t + \Delta t) - F_T(t).$$

If we divide both the left side and the right side by Δt, take the limit as the length of the interval shrinks to zero, and apply the definition of the density function, we get

$$\gamma = \lim_{\Delta t \to 0} \frac{F_T(t + \Delta t) - F_T(t)}{\Delta t} = f_T(t), \qquad \text{for all } t \in (0, L].$$

Now, using the fact that the integral of the density over the full interval of possibilities must be unity,

$$1 = \int_0^L f_T(t)\, dt = \gamma L,$$

we find

$$\gamma = 1/L.$$

Finally, we complete specification of the distribution by noting that values outside the interval $(0, L]$ have 0 probability. Thus

$$f_T(t) = \begin{cases} 1/L, & \text{for } 0 < t \le L \\ 0, & \text{for } t \le 0 \text{ or } t > L \end{cases} \qquad (6.3)$$

and, by integrating f_T,

$$F_T(t) = \begin{cases} 0, & \text{for } t \le 0 \\ t/L, & \text{for } 0 < t \le L \\ 1, & \text{for } t > L. \end{cases} \qquad (6.4)$$

This p.d.f./c.d.f. pair defines *the Uniform distribution on* $(0, L]$, which we have derived from one condition: *the probability that the point of arrival of the observer is within any subinterval of* $(0, L]$ *is proportional to the length of the interval*. A short-hand notation for the Uniform distribution is $\mathcal{U}(0, L]$. This draws attention to the fact that only the interval length is required to derive properties of this distribution.

As an illustration of a variable with a Uniform distribution we can use the time until the next passage of a polar-orbiting satellite, with an orbital period of $L = 101$ min. Let's suppose that you are part of a team that has completed a field experiment; and it is now evident that data from the satellite, which are routinely collected and archived by the National Environmental Satellite Data and Information Service, will be valuable in the analysis of your own observations. Because the use of satellite data was not anticipated, the field experiment was conducted without regard for the orbital schedule. However, now it is necessary to be concerned with the time at which the satellite was directly above the research platform, following and relative to each experimental observation time. The difference between satellite observation time and experimental time is a variable with $\mathcal{U}(0, L]$ distribution.

The Uniform distribution is a particularly simple distribution to work with because its density function is zero everywhere except within a known interval, where it is constant and the c.d.f. is

$$F_T(t) = t/L.$$

Thus, obtaining the correspondences between possible values of T and the cumulative distribution function is elementary. The MINITAB commands "CDF C; UNIFORM 0 L." and "INVCDF C; UNIFORM 0 L." return values for the $\mathcal{U}(0, L]$ c.d.f. and its inverse, for any specified value C or column of values which you have input with the "SET C" command. However, because it is so simple to calculate them directly, you will probably choose to use MINITAB only if you require its neat, tabular presentations.

The basic properties of this distribution are easy to establish, as we show next.

ASSERTION 6.2. *For the Uniform distribution on* $(0, L]$

(a) *The mean is* $L/2$.
(b) *The variance is* $L^2/12$.
(c) *The characteristic function is* $(e^{i\xi L} - 1)/i\xi L$.

Proof. (a)

$$\mu = \mathcal{E}[T] = \int_{-\infty}^{\infty} t\, f_T(t)\, dt = \frac{1}{L}\int_0^L t\, dt = \frac{1}{L}\frac{L^2}{2} = \frac{L}{2}.$$

(b)

$$\sigma^2 = \mathcal{E}[T^2] - \mu^2,$$

where

$$\mathcal{E}[T^2] = \int_{-\infty}^{\infty} t^2 f_T(t)\, dt = \frac{1}{L}\int_0^L t^2\, dt = \frac{1}{L}\frac{L^3}{3} = \frac{L^2}{3}$$

for

$$\sigma^2 = \frac{L^2}{3} - \left(\frac{L}{2}\right)^2 = \frac{L^2}{12}.$$

(c)

$$\Phi_T(\xi) = \int_{-\infty}^{\infty} e^{i\xi t} f_T(t)\, dt$$

$$= \frac{1}{L}\int_0^L e^{i\xi t}\, dt = \frac{1}{L}\left(\frac{1}{i\xi}e^{i\xi t}\Big|_0^L\right) = \frac{e^{i\xi L} - 1}{i\xi L}. \qquad \blacksquare$$

We note that, in this case, it is more difficult to obtain the values of $\mathcal{E}[T^k]$ from the derivatives of $\Phi_T(\xi)$, because of the necessity for making repeated applications of l'Hospital's Rule to obtain the limits as $\xi \to 0$.

Let us return to the example of the times of passage of the orbiting satellite relative to the research observation times of your team project. To establish the distribution of the *shortest* intervals between research and satellite observation times, we define

$$Z = \begin{cases} T, & \text{if } 0 < T \le L/2 \\ (L - T), & \text{if } L/2 < T \le L \end{cases}$$

and derive its distribution from the distribution of T. Recall that T is defined as the waiting time to the following satellite passage. Hence, if $T > L/2$, the shortest interval between research and satellite observation times will be the interval since the previous passage, or $L - T$. Since Z cannot exceed $L/2$, then for any value of $z \le L/2$,

$$\Pr[Z \le z] = \Pr[T \le z \text{ or } (L - T) \le z]$$
$$= \Pr[T \le z \text{ or } T \ge L - z]$$
$$= \Pr[T \le z] + \Pr[T \ge L - z]$$
$$= F_T(z) + 1 - F_T(L - z)$$
$$= z/L + 1 - (L - z)/L = 2z/L.$$

Notice that this may be rewritten and completed as

$$F_Z(z) = \begin{cases} 0, & \text{for } z \le 0 \\ z/(L/2), & \text{for } 0 < z \le (L/2) \\ 1, & \text{for } z > L/2. \end{cases}$$

Since this is the c.d.f. for the Uniform distribution on $(0, L/2]$, then we have established that

$$Z \to \mathscr{U}(0, L/2].$$

From the properties of Uniform distributions we know that Z has mean

$$\mathscr{E}[Z] = (L/2)/2 = L/4$$

and standard deviation which is half the standard deviation of T:

$$\sqrt{\sigma_z^2} = (L/2)^2/12 = L/4\sqrt{3} \approx 0.144L.$$

In the case of the 101-min orbit,

$$\mu_z = 25.25 \text{ min} \qquad \text{and} \qquad \sigma_z = 14.54 \text{ min}.$$

6.4 RANDOM LOCATION DISTRIBUTIONS

So far in this chapter we have considered the temporal aspect of random occurrences. The focus has been on the *time* to the next occurrence of some important phenomenon of the atmosphere or ocean. Now we reverse the roles of time and space, and focus on *distance and area* as variables of interest. By analogy with waiting time distributions we will find (1) the distributions of the distance from an observer to the most proximal occurrence of the phenomenon and of the distance between adjacent occurrences, when we know only the mean density of these events; and (2) the distribution of the area which must be searched for a known event whose specific location is unknown.

In the first case we postulate that *we know the mean number of occurrences of the phenomenon per unit area or volume*, such as would be available from a historical record, and that *we have no advance knowledge that there will be one or more within any specific region*. Think of the region as surrounding the location of an observer. It may be rectangular in shape, like one of the counties of Kansas, or it may be a circular or spherical region, defined by some radial distance from the key location. In any case, we designate an incremental subregion with a small square, as shown in the diagram:

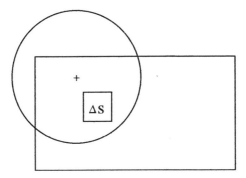

We denote the region by $\mathbf{s}$, its size by $|\mathbf{s}|$, and the regional increment by $\Delta\mathbf{s}$, with size $|\Delta\mathbf{s}|$. Letting λ denote the mean number of occurrences per regional unit, we know from Chapter 5 that

$$Y_{|\mathbf{s}|} = \text{number within region of size } |\mathbf{s}|$$

will have Poisson distribution with parameter λ, provided that the following accurately characterize occurrences of the phenomenon:

(a) The probability of a single occurrence within any regional increment $\Delta\mathbf{s}$ is proportional to $|\Delta\mathbf{s}|$: $p_1(\Delta\mathbf{s}) = \lambda|\Delta\mathbf{s}| + \mathscr{O}_1(|\Delta\mathbf{s}|)$.
(b) The probability of two or more occurrences with $\Delta\mathbf{s}$ is $\mathscr{O}_2(|\Delta\mathbf{s}|)$.
(c) The numbers of occurrences within any non-overlapping subregions are independent of one another.

Thus, when these are a reasonable description of the research setting, the probability mass function for $Y_{|s|}$ is given by

$$p_y(|s|) = \begin{cases} \dfrac{(\lambda|s|)^y}{y!}e^{-\lambda|s|}, & \text{for } y = 0, 1, 2, \ldots \\ 0, & \text{for other real values of } y. \end{cases} \tag{6.5}$$

As we will see, this provides the basis for the distribution of

$R =$ distance to the site of the occurrence nearest the observer.

R is a random distance determined by the random locations of the phenomena under study. Since R will exceed any value r, say, if and only if there are no occurrences within the region of size $|s| = \pi r^2$, then

$$\Pr[R > r] = \Pr[Y_{\pi r^2} = 0] = e^{-\lambda(\pi r^2)}.$$

Accordingly the c.d.f. for R, with $r \geq 0$, is

$$F_R(r) = \Pr[R \leq r] = 1 - \Pr[R > r] = 1 - e^{-\lambda(\pi r^2)}.$$

If $r < 0$, then clearly the probability on the left is 0. Thus the distribution of R is completely characterized by

$$F_R(r) = \begin{cases} 0, & \text{for } r < 0 \\ 1 - e^{-\lambda(\pi r^2)}, & \text{for } r \geq 0 \end{cases} \tag{6.6}$$

with p.d.f.

$$f_R(r) = \begin{cases} 0, & \text{for } r < 0 \\ 2\lambda\pi r\, e^{-\lambda\pi r^2}, & \text{for } r \geq 0. \end{cases} \tag{6.7}$$

With the assumption of the independence of occurrences within nonoverlapping regions, this can also be shown to be the p.d.f. for the distance between adjacent occurrences. (See Exercise 6.12(a).)

We note that this is not an Exponential distribution, because the distance value is squared; and it will have different properties than the distribution of an exponential variable. These can be explored as suggested by Exercise 6.13.

Examples illustrating this distribution are the distance from your research base in Rooks County to the nearest tornado touch down, and the distances between adjacent touch down sites.

We turn now to determination of the distribution of the search area for the site of a known occurrence of a phenomenon, whose specific location within the region under surveillance is unknown. As an examples, let's say that there is a reliable but unidentified report of a lightning strike felling a large tree, within a heavily wooded township, or that a helicopter has inadvertently released a valuable research buoy somewhere along its flight corridor. Here we have situations in which *each increment of the inclusive region may contain the site, equal-sized increments have equal probabilities of containing it, and the probability that the site is within any subregion is proportional to its size.* We will write these in terms of probability statements, using **s** to denote a spatial region subset of some total region **S**, and |s| and |S| to denote their sizes. Thus

$$\Pr[\text{site within } s] = \lambda|s|.$$

And since

$$\Pr[\text{site within } S] = 1$$

then we must have

$$\lambda = 1/|S|.$$

These are the criteria that define the Uniform distribution, here for a spatial domain. In our examples the domain is determined by the total area of the township or the total area of the flight corridor.

Letting Z denote the size of the area searched until the location of the lightning strike or the misplaced buoy is encountered, we have

$$F_Z(z) = \Pr[Z \leq z] = \Pr[\text{site within area of size } z] = \lambda z$$

provided, of course, that $0 < z \leq |S|$. The c.d.f. will clearly be 0 for $z < 0$ and 1 for $z > |S|$. Thus the distribution is completely described by

$$F_Z(z) = \begin{cases} 0, & \text{for } z \leq 0 \\ \lambda z, & \text{for } 0 < z \leq |S| \\ 1, & \text{for } z > |S| \end{cases} \tag{6.8}$$

for which the p.d.f. is

$$f_Z(z) = \begin{cases} 0, & \text{for } z \leq 0 \\ \lambda, & \text{for } 0 < z \leq |S| \\ 0, & \text{for } z > |S|. \end{cases} \tag{6.9}$$

EXERCISES

1. Refer to the lead paragraph of Exercise 11 in Chapter 5.
   (a) Identify the distribution of the time between successive lightning strikes during the electrical storm.
   (b) By direct integration of the density function, find the probability that the interval from the time you start counting to the next strike is: (i) less than 1 min; (ii) between 2 and 4 min; (iii) more than 5 min.
   (c) Using cumulative distribution values generated by MINITAB and the rules for probabilities, confirm the answers you obtained for (b) above.

2. Refer to the example of the tornado watch in Kansas, with $b = 15.3$.
   (a) Find the probability that you must wait at least one week, but not more than three weeks, for the first tornado (i) using the values returned by MINITAB, given in the text; (ii) by integrating the density function for the waiting time distribution.
   (b) Find the time you must plan to wait, say $t_{0.5}$, to have even odds that the first occurrence will not be longer.

3. For an Exponential distribution with parameter λ, prove by direct integration that the mean and variance are

$$\mu = 1/\lambda \qquad \text{and} \qquad \sigma^2 = 1/\lambda^2.$$

4. (a) For the electrical storm of Exercise 1 above, give the mean, standard deviation, and the characteristic function for the time between successive lightning strikes in the initial surveillance section.
   (b) Refer again to Exercise 11 of Chapter 5. Give the mean, standard deviation, and the characteristic function for time between successive lightning strikes in the surveillance area which is 4 times the initial section.

5. (a) In the context of the Kansas tornado watch, evaluate the probability that you will have to wait more than one week for the first tornado sighting.
   (b) Suppose that you have waited a week and there has been no tornado sighting. What is the probability that there will be no tornado sighting in the following week?

6. For the hypothetical Kansas tornado watch, focus on Rooks County. This county has an area of 888 square miles, in the state for which the total area is 82,254 square miles. Assuming that the tornado frequency of $\lambda = 15.3$ per summer, for all of Kansas, is characteristic of Rooks County, find the probability that there will be no tornados in the county during the entire 92-day summer.

7. In the event that the project requires data from 5 tornados, we have found that the expected time interval to complete it is 76.6 days, using the stochastic independence of $T_1, \ldots, T_5$ to get

$$\mathscr{E}\left[\sum_{j=1}^{5} T_j\right] = \sum_{j=1}^{5} \mathscr{E}[T_j].$$

   (a) Use the independence of $T_1, \ldots, T_5$ and the definition

$$\Phi_Z(\xi) = \mathscr{E}[e^{i\xi Z}]$$

   to establish that the characteristic function for $Z = \sum_{j=1}^{5} T_j$ is

$$[\lambda/(\lambda - i\xi)]^5.$$

(b) Based on the fact that the characteristic function for the distribution of a variable uniquely identifies the distribution, answer the question: "Could we assume that the distribution of $T_1 + \cdots + T_5$ is an Exponential distribution?" Support your answer with a solid argument.

8. Refer to the example of the polar-orbiting satellite with orbital period of 101 min and suppose that a current polar-orbiter has the same period. Assuming that you do not have "insider's knowledge" of the schedule for this satellite, find the following.
   (a) The expected waiting time to the next, geographically nearest overpass and the standard deviation for this waiting time.
   (b) The probability that waiting time T is (i) at least 61 min; (ii) less than 40 min; (iii) between 10 and 50 min.

9. Confirm, by use of the characteristic function, that the mean and standard deviation of the waiting time to the next, geographically nearest overpass of the polar-orbiting satellite are as given by Assertion 6.2(a) and (b), and that they are equal to
$$\mu = 50.5 \quad \text{and} \quad \sigma = 29.2 \text{ min}.$$

10. Assuming a fixed orbit, evaluate the probabilities that the time interval to the geographically nearest overpass is: (i) less than 5 min; (ii) at least 10 min; (iii) no more than 20 min; (iv) between 10 and 20 min; (v) exceeding 30 min.
    Hint: This may be either the most recent overpass or the next in time.

11. Imagine that you are involved in a large, ocean-based experiment which has been carefully planned and orchestrated. Preparation included the manufacture of 350 drifting buoys with multiple water-sampling units. These were released by aircraft, at 350 distinct locations over a 10^6 square mile region of the North Atlantic, with release points determined by "a random location generator". The operation of this device guarantees that the probability of a buoy being dropped in any small subregion is proportional to the size of that subregion, and that the numbers in any pair of non-overlapping subregions are independent of one another. Let's say that you are on a ship and that you have the responsibility of finding and retrieving any sighted buoys within a path a quarter of a mile wide.
    (a) If the ship covers 250 straight-line miles over this region on the first day, what is the probability that you will not collect any buoys?
    (b) If the ship covers 250 miles per day for 10 days, what is the probability that you will recover at least two of the buoys?
    (c) Since question (a) pertains to the first day of the field component of the experiment, it is reasonable to assume that the buoy locations are the points at which they were dropped. Since this is

not true for question (b), what implicit assumption must you have made in order to answer this question?

12. Refer to the hypothetical ocean experiment of Exercise 11. Suppose that you have just successfully retrieved one of the buoys. By "probabilistic argument", work out the p.d.f. for the distance to the next nearest buoy for the following:
    (a) in any direction from the pickup point;
    (b) along the direction of the cruise.

13. (a) Find the characteristic function, the mean, and standard deviation for the variable whose distribution is described by expressions (6.6) and (6.7).
    (b) Find the characteristic function, the mean, and standard deviation for the distance to the next nearest buoy, of Exercise 12(a).

14. Return to the example of a reported lightning strike felling a large tree in a heavily wooded township with a total area of 400 square miles. Suppose that a helicopter is coming from outside, starting a site search from one corner.
    (a) What is the expected area its crew will have to search before encountering the site?
    (b) What is the probability they will search at least 324 square miles?

7

THE NORMAL
DISTRIBUTIONS
"good approximations
for many composite variables"

7.1 INTRODUCTION

The Normal distributions are the distributions of many things and of none, in the sense that they provide close approximations, but not precise representations of the distributions of stochastic variables which have many sources of influence. In fact, these distributions are only truly derivable if we can assume there are infinitely many sources of influence and that the contributions from these sources are averaged in determining the value of the variable, at each time and location of observation. In addition, the postulated influences must be assumed to be stochastic inputs which satisfy special conditions on their own low order moments. We never have sufficient knowledge of any observed system to guarantee that these assumption are satisfied; and, of course, we cannot ever identify an infinity of sources of influence.

You may well ask why Normal distributions are so widely used.

The Normal distributions are widely used because, in many circumstances, they can be shown to give good approximations to distributions of descriptive statistics whose true distributions are unknown. Normal distributions have been tabulated and published extensively. Software has been developed which enables researchers to determine whether the distributions of their statistics conform to the best approximating Normal distribution. And additional software facilitates evaluation of research outcomes using this distribution or one of several distributions which may be derived from it. All told, the normal distributions are basic to an extensive and powerful array of data analysis tools. Nonetheless, before we immerse ourselves in their application to the making of scientific inferences from data sets, it is well to be reminded that the outcome will only be as valid as the validity of the Normal approximation to the distribution of the variable whose observed values comprise the data set.

In this chapter and the next, inferences about the central value and the variance of a stochastic variable for which we have a collection of observed values will be constructed from the assumption that this variable has a normal distribution. The inferences are based on the descriptive statistics $\overline{X}$ and S^2, or partitionings of these statistics which enable us to bring the stochastic structures into higher resolution. As we shall see with the development of the topics of Chapters 8 and 9, the Normal distributions and the family of distributions which is derived from them provide keys to many important inference situations. The Normal and Student's t are distributions of an observed mean value, $\overline{X}$, in relation to its true or hypothesized value, μ. The χ^2 and F are distributions of variance statistics in relation to their true or hypothesized values and the composition of the collective variance. Each of these distributions is a powerful inferential tool. And each serves a specific inferential objective, in describing parameters and relationships between parameters that govern the stochastic behavior of an observed variable.

7.2 THE UNIVARIATE NORMAL DISTRIBUTIONS

The Univariate Normal distribution is undoubtedly the most familiar of all distributions to physical scientists. It is continuous and its probability density function is a bell-shaped curve which concentrates two-thirds of the distribution within one standard deviation, 95% within two, and 99% within three standard deviations of its central value. The mean, median, and mode of a normally distributed variable are identical and define the axis of symmetry of the density function. The (only) other explicit parameter of the distribution is the variance which, by its definition, determines the extent of spread of the distribution around is central value.

Designating Univariate Normal Distributions

For a scalar variable X whose distribution is Normal with mean μ and variance σ^2, the probability density function is

$$f(x) = \frac{1}{\sqrt{2\pi\sigma^2}}e^{-(x-\mu)^2/2\sigma^2}, \qquad \text{for } -\infty < x < +\infty. \qquad (7.1)$$

Clearly the p.d.f. is symmetric around μ and its rate of decrease away from μ is dependent on the value of σ^2. Since these are the only parameters in the designation of the distribution, setting values for them completely determines all characteristics of the distribution. With a short computer program, or by using MINITAB, we can generate and plot Univariate Normal p.d.f.s, for varying combinations of parameters, and explore the influence of μ and σ^2 on the shape of the density function. Figure 7.1 provides examples of Normal distribution curves, for several parameter specifications. For comparison, Fig. 7.2 shows histograms of real data. Visual examination suggests that some of the latter could be reasonably approximated by normal curves, calculated by matching the means and variances of the observed data; and some show evidence of properties not shared by any Normal distribution.

Since characteristic functions are a powerful aid to the study of Normal distributions and of distributions of *combinations* of normally distributed variables, it will serve our objectives to establish the following.

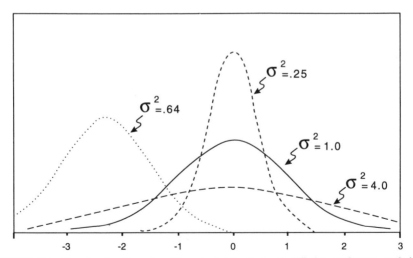

FIGURE 7.1 Density functions for Normal distributions of differing variances, scaled relative to the standard normal curve which is labeled $\sigma^2 = 1.0$.

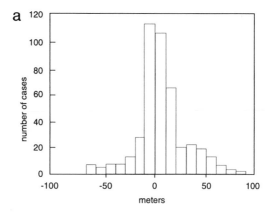

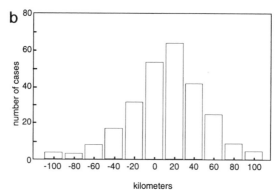

FIGURE 7.2   (a) Histogram of differences in thickness of the isohaline layer and the isothermal layer of the western equatorial Pacific, from 434 profiles (Lukas and Lindstrom, 1991). (b) Histogram of the Gulf Stream axis offset between an altimeter and a real-time Gulfcast derived from crosstrack velocities, for 272 cases during the period from October 7, 1988 to October 4, 1989 (Glen, Porter, and Robinson, 1991).

ASSERTION 7.1.   *The characteristic function for the normal distribution with mean μ and variance σ^2 is*

$$\Phi(\xi) = e^{i\mu\xi - \sigma^2\xi^2/2}. \tag{7.2}$$

Proof.

$$\Phi(\xi) = \int_{-\infty}^{+\infty} e^{i\xi x} \left[\frac{1}{\sqrt{2\pi\sigma^2}} e^{-(x-\mu)^2/2\sigma^2} \right] dx$$

$$= \int_{-\infty}^{+\infty} \frac{1}{\sqrt{2\pi\sigma^2}} e^{-[x^2 - 2(\mu - i\xi\sigma^2)x + \mu^2]/2\sigma^2} dx$$

$$= \left\{ \int_{-\infty}^{+\infty} \frac{1}{\sqrt{2\pi\sigma^2}} e^{-[x-(\mu - i\xi\sigma^2)]^2/2\sigma^2} dx \right\} e^{[(\mu - i\xi\sigma^2)^2 - \mu^2]/2\sigma^2}.$$

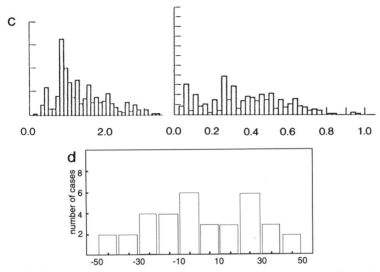

FIGURE 7.2   (c) Histograms of ice draft and snow thicknesses in meters, of Class I floes in the northwestern Weddell Sea, October–November 1988, from measurements made in 1304 drilled holes (Lange and Ericken, 1991). (d) Histogram of residuals from a linear fit to preflight calibration data for the temperature sounder on the GOES satellite. Data provided by the National Environmental Satellite Data Information Service.

The step between the second and third lines is made by completing the square in x, in the exponent, and factoring terms not involving x, out of the integral. If we now substitute $\nu = \mu - i\xi\sigma^2$ and use the fact that

$$\int_{-\infty}^{+\infty} \frac{1}{\sqrt{2\pi\sigma^2}} e^{-(x-\nu)^2/2\sigma^2} \, dx = 1$$

for any ν for which $|\nu|$ is finite, we obtain the result claimed. Specifically,

$$\Phi(\xi) = e^{\{[\mu^2 - 2\mu(i\xi\sigma^2) + (i\xi\sigma^2)^2] - \mu^2\}/2\sigma^2} = e^{i\mu\xi - \xi^2\sigma^2/2}. \qquad \blacksquare$$

Examples illustrating the correspondences between Normal densities and their characteristic functions are presented in Table 7.1.

A shorthand notation which says that variable X has Normal distribution with mean μ and variance σ^2 is $X \to \mathcal{N}(\mu, \sigma^2)$. In some references this is written with the standard deviation σ as the second argument of $\mathcal{N}(\ , \ )$; however, this text will be consistent in its use of the variance. Note that the variance is in squared units of the variable.

Computing Probabilities with a Univariate Normal Distribution

As with any continuous distribution, the probability that variable $X \to \mathcal{N}(\mu, \sigma^2)$ takes a value within any continuous interval $(a, b]$ is the integral

TABLE 7.1 Examples of Densities and Characteristic Functions for Normal Distributions with Parameters μ and σ^2

μ	σ^2	$f(x)$	$\Phi(\xi)$
0	0.25	$0.80e^{-2x^2}$	$e^{-0.125\xi^2}$
0	1.0	$0.40e^{-x^2/2}$	$e^{-0.5\xi^2}$
0	12.5	$0.11e^{-0.04x^2}$	$e^{-6.25\xi^2}$
0.25	0.25	$0.80e^{-(x-0.25)^2}$	$e^{0.25i\xi-0.125\xi^2}$
10	100	$0.04e^{-0.005(x-10)^2}$	$e^{10i\xi-50\xi^2}$

of the p.d.f. over $(a, b]$;

$$\Pr[a < X \le b] = \int_a^b \frac{1}{\sqrt{2\pi\sigma^2}} e^{-(x-\mu)^2/2\sigma^2} dx \qquad (7.3)$$

and the cumulative distribution function, as a special case, is

$$\Pr[X \le x] = \int_{-\infty}^x \frac{1}{\sqrt{2\pi\sigma^2}} e^{-(x'-\mu)^2/2\sigma^2} dx'. \qquad (7.4)$$

Because the Normal distribution is so frequently used and because (7.3) and (7.4) require numerical integration for their evaluation, tables of probabilities are widely available. They appear in the backs of most statistics books and in compilations of mathematical tables. In order to use these tables we must know the values of μ and σ^2, and convert the probability statement about X to a probability statement about the "standardized variable"

$$Z = (X - \mu)/\sigma.$$

It is easy to show that Z has Normal distribution with zero mean and unit variance, for any finite values of μ and σ, provided that the variance is not zero, i.e., $Z \to \mathcal{N}(0, 1)$. Accordingly, only one Normal distribution need be tabulated.

The conversion of probability statements is handled as follows. We rewrite the left-hand sides of (7.4) and (7.3), respectively, as

$$\Pr[X \le x] = \Pr\left[Z = \frac{X-\mu}{\sigma} \le \frac{x-\mu}{\sigma}\right] = F_Z\left(\frac{x-\mu}{\sigma}\right)$$

and

$$\Pr[a < X \le b] = \Pr\left[\frac{a-\mu}{\sigma} < Z \le \frac{b-\mu}{\sigma}\right] = F_Z\left(\frac{b-\mu}{\sigma}\right) - F_Z\left(\frac{a-\mu}{\sigma}\right)$$

and see that the right-hand sides may be evaluated by referring the arguments of F_Z to a table of the *Standardized Normal Distribution*. Once we have established that the distribution of Z is what we claim it to be, we can evaluate any and all probabilities of normally distributed variables for which we know or can hypothesize the means and variances. And we can do this with the distribution of the one variable Z. Thus we return to and prove:

ASSERTION 7.2.   *If* $X \to \mathcal{N}(\mu, \sigma^2)$, *for finite* μ *and* $\sigma^2 > 0$, *then*

$$Z = (X - \mu)/\sigma \to \mathcal{N}(0,1) \quad \text{with } \Phi_Z(\nu) = e^{-\eta^2/2}. \quad (7.5)$$

Proof.   This is most easily proved with characteristic functions, using the properties of expectations of functions of random variables from Chapter 4, together with the known characteristic function for X. Denoting the argument of the ch. fn. for Z for η, we have

$$\Phi_Z(\eta) = \mathcal{E}[e^{i\eta Z}] = \mathcal{E}[e^{i\eta(X-\mu)/\sigma}]$$

$$= \mathcal{E}[e^{i(\eta/\sigma)X - i(\eta/\sigma)\mu}] = \{\mathcal{E}[e^{i(\eta/\sigma)X}]\}e^{-i(\eta/\sigma)\mu}$$

$$= \{\Phi_X(\xi)|_{\xi=\eta/\sigma}\}e^{-i(\eta/\sigma)\mu}$$

$$= \{e^{i(\eta/\sigma)\mu - (\eta/\sigma)^2\sigma^2/2}\}e^{-i(\eta/\sigma)\mu} = e^{-\eta^2/2}.$$

Reference to Table 7.1 confirms that this is the characteristic function for $\mathcal{N}(0,1)$.   ∎

Hence a single table of cumulative probabilities for $\mathcal{N}(0,1)$ will suffice for the evaluation of interval probabilities for all normal distributions with finite means and variances. A condensed table for use with the exercises is provided by Table 7.2.

Generally you will not need to make the conversion to Z or to consult a hard copy table of its probabilities. Statistical software packages will do the work for you, enabling you to phase and answer your questions in terms of the original variable. The MINITAB command sequence that returns the c.d.f. for any specified value of the variable X is

"CDF x; NORMAL μ, σ."

Note that its use requires the input x, μ, and σ. And especially note that

TABLE 7.2   Cumulative Probabilities for the Standardized Normal Distribution $Z \to \mathcal{N}(0,1)$

F	0.50	0.60	0.70	0.75	0.80	0.85	0.90	0.95	0.975
z	0.000	0.253	0.524	0.674	0.841	1.037	1.282	1.645	1.960

Note: The table gives values of z such that $F(z) = \int_{-\infty}^{z} (1/\sqrt{2\pi})e^{-z'^2/2}\,dz'$. The symmetry of the distribution gives us $F(-z) = 1 - F(z)$.

the second parameter identifying the distribution is the standard deviation, σ, *not* the variance.

Another MINITAB command sequence for the normal distribution which you will find extremely useful is

$$\text{"INVCDF } p; \text{ NORMAL } \mu, \sigma.\text{"}$$

This one returns the value of x for which the cumulative probability of the values less than x is p, assuming that $X \rightarrow \mathcal{N}(\mu, \sigma^2)$. You input p, μ, and σ. The capability provided by this command will be valuable in the design of any experiment which requires hypothesis testing, for which you want to identify an interval of extreme values with assigned "significance". As it is programmed, the command identifies an interval on the lower end of the distribution, by its right endpoint, when you input the cumulative probability of the interval. When an interval on the upper end of the distribution is wanted for preselected cumulative probability, write the desired probability as $1 - p$ and input p. For example, if you want the value of x for which all larger values have cumulative probability 0.10, write $1 - p = 0.10$, and input $p = 0.90$ in the call to the INVCDF routine. This will return the x for which the cumulative probability of the values less than x is 0.90, and therefore the probability of all the values to the right of x must be 0.10. Since the Normal distributions are continuous, x itself can be put in either interval without altering the probabilities.

Underlying Assumptions and Evaluation of Normality

Assumptions that lead to appropriate use of the Normal distribution are that *the subject variable is the cumulative expression of influences from many independent sources or components, each of which contributes a stochastic element of finite variance.* If the characteristics of the observed field and the instruments used to obtain data are such that these assumptions are reasonable, then we may anticipate that *the variable is normally distributed.* The is a "low tech" statement of the so-called Central Limit Theorem (CLT). There are a number of mathematically rigorous forms of the CLT in the literature of Statistics. For a discussion of this important theorem in notation similar to that used in this text, see Appendix II of Thiébaux and Pedder (1987).

Whatever your level of knowledge or confidence in assumptions about the sources of variability in your data, it is always good practice to confirm the suitability of the assignment of a Normal distribution to any research variable. Inferential use of a distribution which fails to accurately describe the stochastic behavior of that variable will lead to unsupported conclusions. Because this is a critical issue, we take time and space here to introduce four checks for normality. Two are simple "eyeball checks". The other two are more sophisticated tests which permit the assignment of objective measures of credibility to the assumption that the distribution of the subject variable is Normal. The simplest check for "normality" is a plot of the histogram of the data, such as that returned by MINITAB's "HISTOGRAM C1" command, when you have set the data into Column 1 of a MINITAB file. This is a negative test in the sense that we can see departures from the characteristics of a Normal distribution, such as bimodality, flatness, or skewness, if they are present in the data. It does not provide an objective measure of confidence that a regular-appearing data set is a set of observations of a variable with a Normal distribution. In using this eyeball check you should anticipate that unless your data set is very large, the MINITAB histogram will not have a smooth "bell shape" even when the data *are* a valid Normal sample. This is due to the variability expected within sets of observations of stochastic variables. Figure 7.3 illustrates this point with three histograms constructed from Normal data sets, with $N = 25, 100, 500$ observations.

Another visual check of the normality of a data set is provided by the MINITAB command sequence "NSCORES C1 C10. PLOT C1 VS C10.", with the data in column 1. The plot will be approximately a straight line if the distribution of the data is reasonably approximated by a Normal distribution. Again, this is a check which is most useful for illustrating strong departures from normality, rather than providing positive validation. Figure 7.4 presents the NSCORE plots for the data which produced Fig. 7.3. As can be seen here, the lines are not strictly straight lines even for data known to be obtained from a true Normal distribution. So, again, the judgment of whether a Normal distribution provides a satisfactory approximation is a subjective judgment.

Two tests which provide quantitative measures of the reasonableness of the choice of a Normal distribution for a subject variable are tests developed by Lilliefors (see Conover, 1980, pp. 357–363) and by D'Agostino and Pearson. (See D'Agostino *et al.*, 1990, pp. 316–321.) The Lilliefors test for normality is based on a well known test which you may know by the name "Kolmogorov". It differs from the Kolmogorov test in not requiring prior knowledge of the true mean and variance of the distribution. Thus it is more suitable for our purposes, since our information is generally limited to what can be obtained from a data set. The procedure compares the sample cumulative distribution, which is constructed from observed values $x_1, \ldots, x_N$ as

$$G(x) = [\text{number of values which are } \leq x]/N$$

with the c.d.f. of the hypothetical Normal distribution. The sample cumulative distribution (s.c.d.) is always a step function:

(i) zero to the left of the smallest observed value,

(ii) increasing by steps of $1/N$ at successive values of the numerically ordered observations, and constant between successive values,

(iii) equal to 1.0 for all values to the right of the largest observed value.

a Histogram of C1   N = 25

```
Midpoint    Count
    -1.6      2    **
    -1.2      1    *
    -0.8      3    ***
    -0.4      5    *****
     0.0      5    *****
     0.4      6    ******
     0.8      1    *
     1.2      2    **
```

b Histogram of C3   N = 100

```
Midpoint    Count
    -3.0      1    *
    -2.5      0
    -2.0      2    **
    -1.5      7    *******
    -1.0     15    ***************
    -0.5     15    ***************
     0.0     18    ******************
     0.5     17    *****************
     1.0     12    ************
     1.5      6    ******
     2.0      6    ******
     2.5      1    *
```

c Histogram of C4   N = 500
Each * represents 5 obs.

```
Midpoint    Count
    -3.0      1    *
    -2.5      2    *
    -2.0     14    ***
    -1.5     32    *******
    -1.0     56    ************
    -0.5     82    *****************
     0.0     86    ******************
     0.5    108    **********************
     1.0     57    ************
     1.5     38    ********
     2.0     18    ****
     2.5      6    **
```

FIGURE 7.3 Histograms generated by random sampling from a Normal distribution with $\mu = 0.0$ and $\sigma^2 = 1.0$, (a) with $N = 25$, (b) with $N = 100$, and (c) with $N = 500$.

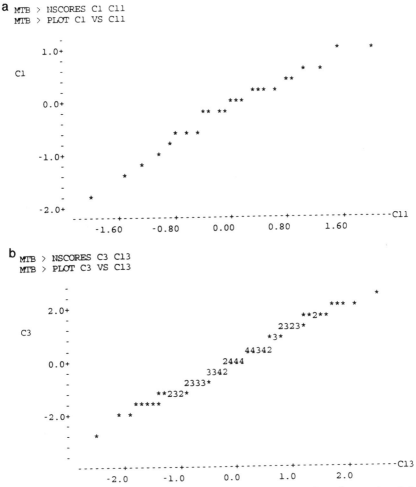

FIGURE 7.4 (a) and (b) Plots of the "normal scores" for the values that produced the histograms of Fig. 7.3, with $N = 25$ and with $N = 100$.

Since the c.d.f. for a normal distribution is a continuous curve, increasing from 0 at $-\infty$, to 1 at $+\infty$, the comparison will be analogous to that shown in Fig. 7.5. The test for normality uses the maximum difference between the curves relative to the number of observations. Thus the statistic evaluated for the determination of normality is

$$\text{MD} = \max_{x} \left| F^{*}(x) - G(x) \right|$$

in which F^{*} denotes the c.d.f. of the hypothetical Normal distribution with

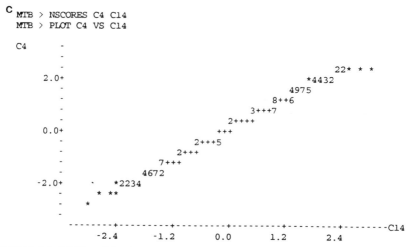

```
c
 MTB > NSCORES C4 C14
 MTB > PLOT C4 VS C14

 C4      -
         -
         -
         -                                                           22*  *  *
    2.0+ -                                                 *4432
         -                                              4975
         -                                           8++6
         -                                        3+++7
         -                                     2++++
    0.0+ -                                    +++
         -                               2+++5
         -                            2+++
         -                        7+++
         -                     4672
   -2.0+ -         ·       *2234
         -           *  **
         -         *
         -
            --------+---------+---------+---------+---------+--------C14
              -2.4      -1.2       0.0       1.2       2.4
```

FIGURE 7.4 (c) Plot of the "normal scores" for the values that produced the histogram of Fig. 7.3 (c) with $N = 500$.

mean and variance equal to $\bar{x}$ and s^2. Close agreement between the sample cumulative distribution and the Normal c.d.f. with the same mean and variance will be reflected by a small value of the maximum difference (MD), and lack of agreement by a large value of this statistic. Probabilities associated with large values of MD, based on the assumption that they would be observed by chance with observations from a true Normal distribution, are given in Table 7.3.

When the number of observations exceeds 30, a good assessment of normality is provided by a MINITAB macro.[1] With this software, the test for normality can be executed for any data in a MINITAB file, with a simple command. The software package contains options for doing separate tests for departures from normality due to suspected skewness or relative flatness of the distribution of the research variable, or doing a single test which is sensitive to departures from normality by either of these criteria. Because the options are *tests of hypotheses* that the distribution of an observed variable is a Normal distribution, they will be discussed more thoroughly in the chapter on hypothesis testing.

The principal keys to tests for normality are two descriptive statistics. The first is a measure of the asymmetry or "skewness" of the array of values of the data set

$$\sum_{j=1}^{N} \left(x_j - \bar{x} \right)^3 \Bigg/ \left[\sum_{j=1}^{N} \left(x_j - \bar{x} \right)^2 \right]^{3/2}$$

---

[1]This is available by request from MINITAB Inc., 3081 Enterprise Drive, State College, PA 16801-2756.

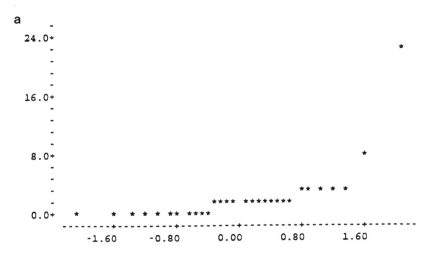

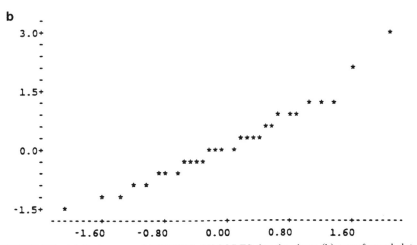

FIGURE 7.5 (a) Data versus MINITAB's NSCORES for the data; (b) transformed data versus NSCORES for the transformed data.

and the other is a measure of its flatness or "kurtosis"

$$\sum_{j=1}^{N} \left(x_j - \bar{x} \right)^4 \bigg/ \left[\sum_{j=1}^{N} \left(x_j - \bar{x} \right)^2 \right]^2 .$$

The D'Agostino–Pearson K^2 statistic is sensitive to departures from normality in either of these two parameters. For a Normal distribution, the expected values of the skewness and the so-called kurtosis statistics are

TABLE 7.3   Critical Values for Lillifors Test for Normality: $\Pr[\max F^*(x) - G(x) > z \mid X \to \mathcal{N}(\mu, \sigma^2)]$, Where N = Number of Observations Going into the Estimate $G(x)$

N	\(\alpha\) 0.10	0.05	0.02	0.01	N	\(\alpha\) 0.10	0.05	0.02	0.01
1	—	—	—	—	16	0.1954	0.2129	0.2332	0.2476
2	—	—	—	—	17	0.1901	0.2071	0.2270	0.2410
3	0.3666	0.3758	0.3812	0.3830	18	0.1852	0.2017	0.2212	0.2349
4	0.3453	0.3753	0.4007	0.4131	19	0.1807	0.1968	0.2158	0.2292
5	0.3189	0.3431	0.3755	0.3970	20	0.1765	0.1921	0.2107	0.2238
6	0.2972	0.3234	0.3523	0.3708	21	0.1725	0.1878	0.2060	0.2188
7	0.2802	0.3043	0.3321	0.3509	22	0.1688	0.1838	0.2015	0.2141
8	0.2652	0.2880	0.3150	0.3332	23	0.1653	0.1800	0.1974	0.2097
9	0.2523	0.2741	0.2999	0.3174	24	0.1620	0.1764	0.1936	0.2056
10	0.2411	0.2619	0.2869	0.3037	25	0.1589	0.1730	0.1899	0.2018
11	0.2312	0.2514	0.2754	0.2916	26	0.1560	0.1699	0.1865	0.1981
12	0.2225	0.2420	0.2651	0.2810	27	0.1533	0.1670	0.1833	0.1947
13	0.2148	0.2336	0.2559	0.2714	28	0.1507	0.1642	0.1802	0.1915
14	0.2077	0.2261	0.2476	0.2627	29	0.1483	0.1615	0.1773	0.1884
15	0.2013	0.2192	0.2401	0.2549	30	0.1460	0.1589	0.1746	0.1855

From Neave and Worthington (1988, p. 374).
Note: The numbers in the body of the table are the values of z for the corresponding values of α and N.

0 and 3, respectively. The D'Agostino–Belanger–D'Agostino package evaluates the probabilities that a true Normal distribution could produce values as large as those of the given data set. These probabilities may then be interpreted as measures of credibility of the normality assumption.

Transforming Data to Produce Normality

If we must conclude, on the basis of any of the foregoing tests, that a Normal distribution is not a suitable description of the stochastic behavior of the observed variable, what options are open to us? The first possibility we may wish to consider is that a simple transformation of the variable, such as its logarithm or its inverse, may produce a variable whose distribution is very close to a Normal distribution. In this case, the transformed variable may be used, with appropriate restructuring of the inferences. Statistical software packages make it easy to explore variable transformation options and check their results for normality.

A good example of a failed check for normality and a successful variable transformation is shown in Figs. 7.5(a) and 7.5(b). The first is a plot of the original data versus the NSCORES for the data. The second is a plot produced following a logarithmic transformation of the original data, with NSCORES for the transformed data. These results lend confidence to basing analyses on an assumed Normal distribution for the transformed variable. When a suitable transformation is found, the inferences required by your research program may then be rephrased in terms of the new variable for which you *can* assume a Normal distribution. Exercises 7.6 and 7.7 illustrate the possibilities for detecting non-normality and carrying out "corrective" transformations.

7.3 MULTIVARIATE NORMAL DISTRIBUTIONS

Unlike the situations in which we use the binomial and Poisson distributions of enumeration variables, or the Exponential and Uniform distributions for time and location variables, we virtually never use a normal distribution with a single scalar value of a variable in scientific research. A normal distribution is generally applied when we have observed many values of a variable, perhaps at several times or locations, and frequently when we have measurements of distinct variables. For example, the latter might be the components of wind or current velocity, together with pressure or density, all of which may have been recorded for several locations. To handle these we need to extend our notation and our concepts of properties of distributions to variables with several scalar components. For the Normal distribution this is easily done with vector and matrix notation.

Designating Multivariate Normal Distributions

Let's begin with a distribution for two variables considered together, say wind or current velocity components. We will assume that they have a joint Bivariate Normal distribution.[2] The mental image associated with a Bivariate Normal distribution is a three-dimensional bell sitting on the (X,Y) plane. The height of the surface of the bell above point (x, y) is the corresponding value of the joint probability density function. To write the general formulation of the Bivariate Normal p.d.f., we denote the means for the variables X and Y by μ and ν, their variances by σ^2 and τ^2, respectively, and their correlation by ρ. In terms of these parameters,

---

[2] We have chosen velocity components to illustrate the Bivariate Normal distribution because we know that they are not independent; and we wish to draw attention early to the fact that joint normality makes no assumption of independence of the component variables.

the density function is

$$f_{X,Y}(x,y) = \frac{1}{2\pi\sqrt{\sigma^2\tau^2(1-\rho^2)}}e^{-[(x-\mu)^2/\sigma^2-2(x-\mu)(y-\nu)\rho/\sigma\tau+(y-\nu)^2/\tau^2]/2(1-\rho^2)}$$

for all pairs of values (x, y) in the plane. Alternatively, we may write

$$f_{X,Y}(x,y) = \left[(2\pi)^2|\Sigma|\right]^{-1/2} e^{-(1/2)\left(\begin{smallmatrix}x-\mu\\y-\nu\end{smallmatrix}\right)^{\top}\Sigma^{-1}\left(\begin{smallmatrix}x-\mu\\y-\nu\end{smallmatrix}\right)}, \qquad (7.6)$$

where

$$\Sigma = \begin{pmatrix} \sigma^2 & \rho\sigma\tau \\ \rho\sigma\tau & \tau^2 \end{pmatrix} \qquad (7.7)$$

is the matrix of variances and covariances of X and Y, and Σ^{-1} is its matrix inverse. The notation $(\ )^{\top}$ indicates the transpose of a column vector.

Recall that a density function does not assign probabilities to the values of the variables which are written into its formulation. For a *univariate* distribution, the *area* under the density function, over a given interval of possible values for the variable, is the probability that the variable is observed to have one of these values. For a *bivariate* distribution, it is the *volume* between the density function surface and a region selected in the (X,Y) plane, that equals the probability of the pair of variables taking a value in that region. Thus, again, the probability of observing a pair of values within any designated (continuous) region could be evaluated if we could integrate the bivariate density over the region. It is important to hold this concept in mind, even though we will never be faced with trying to do this for any of the Normal distributions.

It is simple to extend (7.6) and (7.7) to describe the collective distribution of N variables. We represent the variables as the components of a stochastic vector array with its possible values in Euclidean N-space,

$$\mathbf{X} = \begin{pmatrix} X_1 \\ \vdots \\ X_N \end{pmatrix} \qquad \text{and} \qquad \mathbf{x} = \begin{pmatrix} x_1 \\ \vdots \\ x_N \end{pmatrix}$$

and denote the vector array of corresponding mean values and the $N \times N$ matrix of variances and covariances by

$$\boldsymbol{\mu} = \begin{pmatrix} \mu_1 \\ \vdots \\ \mu_N \end{pmatrix} \qquad \text{and} \qquad \Sigma = \left(\mathscr{E}\left[(x_j - \mu_j)(x_k - \mu_k)\right]\right) = (\sigma_{jk}). \quad (7.8)$$

With these designations, *the density function for the Multivariate Normal distribution of* $X_1, \ldots, X_N$ is

$$f_{\mathbf{X}}(\mathbf{x}) = \left[(2\pi)^N |\ddagger| \right]^{-1/2} e^{-(1/2)(\mathbf{x}-\boldsymbol{\mu})^\top \Sigma^{-1}(\mathbf{x}-\boldsymbol{\mu})}, \qquad \text{for all } \mathbf{x} \in \mathscr{R}^N. \quad (7.9)$$

The probability that coincident observation of the variables results in an array of values that satisfies $a_j < x_j \leq b_j$ for *all* of $j = 1, \ldots, N$, conceptually, is the integral of the joint density function over that region of $\mathscr{R}^N$,

$$\Pr[\, a_1 < X_1 \leq b_1, \ldots, a_N \leq X_N \leq b_N\,]$$
$$= \int_{a_N}^{b_N} \cdots \int_{a_1}^{b_1} \left[(2\pi)^N |\Sigma| \right]^{-1/2} e^{-(1/2)(\mathbf{x}'-\boldsymbol{\mu})^\top \Sigma^{-1}(\mathbf{x}'-\boldsymbol{\mu})} \, dx'_1 \ldots dx'_N.$$
$$(7.10)$$

If you can visualize things in many dimensions, you can think of this as the volume defined by the density function surface and a continuous "rectangular" region of an N-dimensional hyperplane. In any case, we will be using a multidimensional analog of a Bivariate Normal distribution.

The cumulative distribution function is a special case of (7.10). Specifically it is the probability of the simultaneous events $X_j \leq x_j$, for $j = 1, \ldots, N$, which we write as

$$F_{\mathbf{X}}(\mathbf{x}) = \Pr[\, X_1 \leq x_1, \ldots, X_N \leq x_N\,]$$
$$= \int_{-\infty}^{x_N} \cdots \int_{-\infty}^{x_1} \left[(2\pi)^N |\ddagger| \right]^{1/2}$$
$$\times e^{-(1/2)(\mathbf{x}'-\boldsymbol{\mu})^\top \Sigma^{-1}(\mathbf{x}'-\boldsymbol{\mu})} \, dx'_1 \ldots dx'_N. \quad (7.11)$$

When the distribution of a stochastic vector $\mathbf{X}$ is accurately described by the Multivariate Normal distribution with mean vector $\boldsymbol{\mu}$ and variance/ covariance matrix $\ddagger$, we will write

$$\mathbf{X} \to \mathscr{N}(\boldsymbol{\mu}, \Sigma).$$

In scientific research the Multivariate Normal distribution is seldom used explicitly because its use would require either

(a) knowledge of $\boldsymbol{\mu}$ and hypothesized values for the elements of $\ddagger$, or

(b) knowledge of $\ddagger$ and hypothesized values for the elements of $\boldsymbol{\mu}$.

Usually we are not certain of either $\boldsymbol{\mu}$ or $\ddagger$. We are constrained to estimate the elements of both $\boldsymbol{\mu}$ and $\ddagger$ from the information in a data set. In these situations, we use one of the distributions which has been derived from the Multivariate Normal distribution to utilize the level of information at hand. The structure of the Multivariate Normal distribution is presented and studied in some detail because the properties and utilities of the distributions derived from it (which do most of the work in statistical data analysis) are best understood in relation to this structure.

A central component of the Multivariate Normal distribution is the $N \times N$ matrix of variances and covariances in (7.8) which we will also write as

$$\Sigma = \mathscr{E}\left[(\mathbf{X} - \boldsymbol{\mu})(\mathbf{X} - \boldsymbol{\mu})^{\top}\right]. \qquad (7.12)$$

This matrix will always be *symmetric*, regardless of the relationships of the individual elements of $\mathbf{X}$ and their joint stochastic structure. If we assume, as we will throughout our discussions, that all N elements are random variables with positive variances and that none is redundant in the sense that its value could always be obtained as a fixed linear combination of any others, then Σ will also be *positive definite*. These two properties, positive variance and non-redundant elements of the data set, are critical to analysis of any collection of data and should be confirmed at the outset. A "redundant" variable can be inadvertently included in a multivariate analysis, with consequences that would be difficult to decipher. For example, in haste we might decide to analyze the peak wind speeds in the CLIMAT data file, and put the other eight variables into the hopper together, as "predictors." That would put daily average temperature together with daily maximum and minimum temperatures, from which it is computed as the average. Thus the matrix of variances and covariances would be singular; and the use of its inverse in any analysis would produce invalid results. Hopefully the results would be sufficiently bizarre to signal an error; although there is no guarantee that they will be. The possibility of getting totally misleading, but unsuspected analysis outcomes, is one of the hazards of the computer age. Generally data sets are too large for critical, visual examination to be practical. It is essential to carry out preliminary diagnostics electronically, to be certain that data characteristics match the requirements for analysis. (You will discover that time spent in this way can be an enormous saving in the long run!)

Any symmetric, positive definite matrix may be factored into the product of a nonsingular matrix times its transpose. Consequently we may write

$$\Sigma = \mathbf{B}\mathbf{B}^{\top} \qquad \text{and} \qquad \Sigma^{1/2} = \mathbf{B},$$

where $\mathbf{B}$ is nonsingular. These relationships and the definition of $\mathbf{A} = \mathbf{B}^{-1}$, which satisfy

$$\mathbf{B}\mathbf{B}^{\top} = \Sigma, \qquad \mathbf{A}^{\top}\mathbf{A} = \Sigma^{-1}, \qquad \text{and} \qquad \mathbf{A}\mathbf{B} = \mathbf{I} = B^{\top}A^{\top},$$

will be useful in the proof of Assertion 7.3 and subsequently in identifying transformations which "standardize" vector variables.

We now turn to the characteristic function for a Multivariate Normal distribution. We give a thorough proof of the form claimed for the characteristic function, because it provides insight to some powerful techniques for data analysis.

ASSERTION 7.3. *If* $\mathbf{X} \rightarrow \mathcal{N}(\boldsymbol{\mu}, \Sigma)$, *then*

$$\Phi_{\mathbf{X}}(\boldsymbol{\xi}) = e^{i\boldsymbol{\xi}^{\top}\boldsymbol{\mu} - (1/2)\boldsymbol{\xi}^{\top}\Sigma\boldsymbol{\xi}}.$$

Proof. If we assume $\mathbf{X} \rightarrow \mathcal{N}(\boldsymbol{\mu}, \Sigma)$, then by definition of the characteristic function

$$\Phi_{\mathbf{X}}(\boldsymbol{\xi}) = \mathcal{E}\left[e^{i\boldsymbol{\xi}^{\top}\mathbf{x}}\right] = \int_{\mathcal{R}^N} e^{i\boldsymbol{\xi}^{\top}\mathbf{x}} \left[(2\pi)^N |\Sigma|\right]^{-1/2} e^{-(1/2)(\mathbf{x}-\boldsymbol{\mu})^{\top}\Sigma^{-1}(\mathbf{x}-\boldsymbol{\mu})} \, d\mathbf{x}.$$

Now, if we make a linear transformation to a new integration variable,

$$\mathbf{z} = \Sigma^{-1/2}(\mathbf{x} - \boldsymbol{\mu}) = \mathbf{A}(\mathbf{x} - \boldsymbol{\mu}),$$

where $\mathbf{A} = \Sigma^{-1/2}$, we may rewrite the integral above in terms of $\mathbf{z}$ and the Jacobian of the transformation (Graybill, 1983, p. 328):

$$J = \left|\frac{d\mathbf{x}}{d\mathbf{z}}\right| = |\Sigma^{1/2}|.$$

Thus

$$(\mathbf{x} - \boldsymbol{\mu}) = \mathbf{A}^{-1}\mathbf{z} = \mathbf{B}\mathbf{z} \quad \text{and} \quad \mathbf{x} = \mathbf{A}^{-1}\mathbf{z} + \boldsymbol{\mu} = \mathbf{B}\mathbf{z} + \boldsymbol{\mu}$$

which enable us to write

$$\Phi_{\mathbf{X}}(\boldsymbol{\xi}) = \int_{\mathcal{R}^N} e^{i\boldsymbol{\xi}^{\top}(\mathbf{B}\mathbf{z}+\boldsymbol{\mu})} \left\{\left[(2\pi)^N |\Sigma|\right]^{-1/2} e^{-(1/2)(\mathbf{B}\mathbf{z})^{\top} A^{\top} A(\mathbf{B}\mathbf{z})}\right\} |\Sigma|^{1/2} \, d\mathbf{z}$$

$$= e^{i\boldsymbol{\xi}^{\top}\boldsymbol{\mu}} \int_{\mathcal{R}^N} (2\pi)^{-N/2} e^{-(1/2)[\mathbf{z}^{\top}\mathbf{z} - 2i\boldsymbol{\xi}^{\top}\mathbf{B}\mathbf{z}]} \, d\mathbf{z}.$$

If we now substitute $\boldsymbol{\eta}$ for $\mathbf{B}^{\top}\boldsymbol{\xi}$ and complete the product within the square brackets, above, we have

$$\mathbf{z}^{\top}\mathbf{z} - 2i\boldsymbol{\xi}^{\top}\mathbf{B}\mathbf{z} = (\mathbf{z}^{\top}\mathbf{z} - 2i\boldsymbol{\eta}^{\top}\boldsymbol{\eta} - \boldsymbol{\eta}^{\top}\boldsymbol{\eta}) + \boldsymbol{\eta}^{\top}\boldsymbol{\eta} = (\mathbf{z} - i\boldsymbol{\eta})^{\top}(\mathbf{z} - i\boldsymbol{\eta}) + \boldsymbol{\eta}^{\top}\boldsymbol{\eta}$$

for

$$\Phi_{\mathbf{x}}(\boldsymbol{\xi}) = e^{i\boldsymbol{\xi}^{\top}\boldsymbol{\mu}} \left[\int_{\mathcal{R}^N} (2\pi)^{-N/2} e^{-(1/2)(\mathbf{z}-i\boldsymbol{\eta})^{\top}(\mathbf{z}-i\boldsymbol{\eta})} \, d\mathbf{z}\right] e^{-(1/2)\boldsymbol{\eta}^{\top}\boldsymbol{\eta}}.$$

Finally we use the fact that

$$\int_{\mathcal{R}^N} (2\pi)^{-N/2} e^{-(1/2)(\mathbf{z}-i\boldsymbol{\eta})^{\top}(\mathbf{z}-i\boldsymbol{\eta})} \, d\mathbf{z} = 1 \qquad \text{for } |i\boldsymbol{\eta}| < \infty$$

and complete the proof as

$$\Phi_X(\xi) = e^{i\xi^T\mu}e^{-(1/2)\eta^T\eta}\big|_{\eta=B^T\xi}$$

$$= e^{i\xi^T\mu-(1/2)\xi^T(BB^T)\xi} = e^{i\xi^T\mu-(1/2)\xi^T\Sigma\xi}. \qquad \blacksquare$$

VIP: We note that since the characteristic function is the Fourier Transform of the density function

$$\Phi_X(\xi) = \int e^{i\xi^T x} f(x)\, dx$$

with inverse

$$f_X(x) = \int e^{-i\xi^T x} \Phi_X(\xi)\, d\xi$$

then the implication of the claim goes both ways. We may write the corollary as

$$f_X(x) = \left[(2\pi)^N |\Sigma|\right]^{-1/2} e^{-(1/2(x-\mu)^T\Sigma^{-1}(x-\mu)} \Leftrightarrow$$

$$\Phi_X(\xi) = e^{i\xi^T\mu-(1/2)\xi^T\Sigma\xi}. \qquad (7.13)$$

The Distributions of Linear Transformations of Normally Distributed Vectors

The transformation to the variable

$$Z = \Sigma^{-1/2}(X-\mu) = A(X-\mu),$$

where $A = B^{-1}$, and B is from the factorization $\Sigma = BB^T$, gives us an extremely powerful tool which is used in many contexts.

ASSERTION 7.4. $X \to \mathcal{N}(\mu, \Sigma) \Rightarrow Z \to \mathcal{N}(0, I)$, where I is the $N \times N$ identity matrix with ones on the diagonal and zeros everywhere else.

Proof.   Again we let $A = B^{-1}$ and recall that

$$AB = I = B^T A^T \qquad \text{and} \qquad A(BB^T)A^T = I.$$

Now

$$Z = \Sigma^{-1/2}(X-\mu) = A(X-\mu)$$

has mean

$$\mathcal{E}[Z] = \mathcal{E}[A(X-\mu)] = A\,\mathcal{E}[X-\mu] = 0,$$

variance/covariance matrix

$$\mathscr{E}[\mathbf{Z}\mathbf{Z}^\top] = \mathscr{E}\{[A(\mathbf{X} - \boldsymbol{\mu})][A(\mathbf{X} - \boldsymbol{\mu})]^\top\}$$

$$= \mathscr{E}\{A(\mathbf{X} - \boldsymbol{\mu})(\mathbf{X} - \boldsymbol{\mu})^\top A^\top\}$$

$$= A\mathscr{E}\{(\mathbf{X} - \boldsymbol{\mu})(\mathbf{X} - \boldsymbol{\mu})^\top\}A^\top$$

$$= A\maltese A^\top = A(BB^\top)A^\top = I$$

and characteristic function

$$\Phi_\mathbf{Z}(\boldsymbol{\xi}) = \mathscr{E}\left[e^{i\boldsymbol{\xi}^\top \mathbf{Z}}\right] = \mathscr{E}\left[e^{i\boldsymbol{\xi}^\top A(\mathbf{X} - \boldsymbol{\mu})}\right]$$

$$= \mathscr{E}\left[e^{i(A^\top \boldsymbol{\xi})^\top \mathbf{X}}\right]e^{-i\boldsymbol{\xi}^\top A\boldsymbol{\mu}} = \Phi_\mathbf{X}(A^\top \boldsymbol{\xi})\; e^{-i\boldsymbol{\xi}^\top A\boldsymbol{\mu}}$$

$$= e^{i(A^\top \boldsymbol{\xi})^\top \boldsymbol{\mu} - (1/2)(A^\top \boldsymbol{\xi})^\top \Sigma(A^\top \boldsymbol{\xi})}e^{-i\boldsymbol{\xi}^\top A\boldsymbol{\mu}}$$

$$= e^{-(1/2)\boldsymbol{\xi}^\top[A(BB^\top)A^\top]\boldsymbol{\xi}}$$

$$= e^{-(1/2)\boldsymbol{\xi}^\top \boldsymbol{\xi}}.$$

We may confirm that this is the characteristic function for $\mathscr{N}(\mathbf{0}, I)$ by substituting $\boldsymbol{\mu} = \mathbf{0}$ and $\maltese = I$ into (7.13). ■

VIP: We have just established a result of major significance; namely that whatever the mean vector and variance/covariance matrix of the vector variable X, there is a linear transformation to a vector variable

$$\mathbf{Z} \to \mathscr{N}\left(\begin{pmatrix} 0 \\ \vdots \\ 0 \end{pmatrix}, \begin{pmatrix} 1 & & 0 \\ & \ddots & \\ 0 & & 1 \end{pmatrix}\right).$$

That is, each of the components of $\mathbf{Z}$ has mean 0 and variance 1, and the components are uncorrelated.

We will also have use of a more general result, which we now establish.

ASSERTION 7.5. *If* $\mathbf{X} \to \mathscr{N}(\boldsymbol{\mu}, \maltese)$ *and we define* $\mathbf{Y} = Q\mathbf{X}$, *then*

$$\mathbf{Y} \to \mathscr{N}\left(Q\boldsymbol{\mu}, Q\maltese Q^\top\right) \tag{7.14}$$

for any linear transformation with $M \times N$ *matrix* Q *such that* $Q\Sigma Q^\top$ *is nonsingular.*

Proof. As in the foregoing discussion, we assume here that $\mathbf{X}$ and $\boldsymbol{\mu}$ are $N \times 1$ vectors, and that the $N \times N$ matrix $\maltese$ is nonsingular. By using the characteristic function together with (7.13), the proof is extremely simple:

$$\Phi_Y(\xi) = \mathcal{E}\left[e^{i\xi^\top(Q\mathbf{X})}\right] = \mathcal{E}\left[e^{i(Q^\top\xi)^\top\mathbf{X}}\right] = \Phi_X(Q^\top\xi)$$

$$= e^{i(Q^\top\xi)^\top\mu - (1/2)(Q^\top\xi)^\top \ddagger(Q^\top\xi)}$$

$$= e^{i\xi^\top(Q\mu) - (1/2)\xi^\top(Q\Sigma Q^\top)\xi}.$$

The latter is just the characteristic function of (7.13), with $M \times 1$ mean vector $Q\mu$ and $M \times M$ variance/covariance matrix $Q\ddagger Q^\top$.   ∎

VIP: We can obtain the distribution of the arithmetic average of the components of $\mathbf{X}$ as a special case of this result. We write

$$\bar{X} = \sum_{j=1}^{N} X_j/N = Q\mathbf{X}, \qquad \text{where } Q = \left(\frac{1}{N} \cdots \frac{1}{N}\right)$$

and use (7.14) to get

$$\bar{X} \to \mathcal{N}\left(Q\mu, Q\ddagger Q^\top\right),$$

where

$$Q\mu = \sum_{j=1}^{N} \mu_j/N \qquad \text{and} \qquad Q\ddagger Q^\top = \sum_{j,k} \sigma_{j,k}/N^2.$$

In other words, the sample mean of N normally distributed variables is itself normally distributed, with (scalar) mean and variance

$$\mu_{\bar{X}} = \sum_{j=1}^{N} \mu_j/N \qquad \text{and} \qquad \sigma_{\bar{X}}^2 = \sum_{j,k} \sigma_{jk}/N^2.$$

In the special circumstances when $\ddagger$ is diagonal, i.e., the components of $\mathbf{X}$ are independent of one another, with variances $\sigma_1^2, \ldots, \sigma_N^2$, and covariances all equal to zero, the variance of the sample mean is N^{-2} times the sum of the component variances

$$\sigma_{\bar{X}}^2 = \sum_{j=1}^{N} \sigma_j^2/N^2.$$

If the components have common variance σ^2, then

$$\sigma_{\bar{X}}^2 = N\sigma^2/N^2 = \sigma^2/N.$$

The Multivariate Normal Distribution with Independent Components

The fact that diagonality of Σ guarantees independence of the components of $\mathbf{X}$ is a uniquely important property of the Multivariate Normal distribution and is established by the proof of Assertion 7.6, below. As we noted in Chapter 3, in general $\rho = 0$ and $\text{Cov}(X, Y) = 0$ do not guarantee independence of the variables.

ASSERTION 7.6.   *For* $\mathbf{X} \rightarrow \mathcal{N}(\boldsymbol{\mu}, \Sigma)$ *with*

$$\Sigma = \begin{pmatrix} \sigma_1^2 & & 0 \\ & \ddots & \\ 0 & & \sigma_N^2 \end{pmatrix} \qquad (7.15)$$

the density function for the joint distribution of $X_1, \ldots, X_N$ *factors into the product of the densities for the marginal distributions of the* $X_1, \ldots, X_N$.

Proof.   We will first establish that the conditions of the assertion imply that the marginal distribution of X_j is normally distributed with mean μ_j and variance σ_j^2. Recall that the characteristic function for one of several variables, such as X_j among $X_1, \ldots, X_N$, may be found from the characteristic function for their joint distribution by setting all elements of $\boldsymbol{\xi}$ except ξ_j equal to 0. With the Multivariate Normal distribution, this gives us

$$\Phi_{\mathbf{X}}(\boldsymbol{\xi})\Big|_{\xi_k = 0,\, k \neq j} = e^{i\boldsymbol{\xi}^\top \boldsymbol{\mu} - (1/2)\boldsymbol{\xi}^\top \Sigma \boldsymbol{\xi}}\Big|_{\xi_k = 0,\, k \neq j} = e^{i\xi_j \mu_j - (1/2)\xi_j^2 \sigma_j^2}.$$

The expression on the far right is the ch. fn. for the Univariate Normal distribution. Since j was arbitrary, then we have $X_j \rightarrow \mathcal{N}(\mu_j, \sigma_j^2)$, for $j = 1, \ldots, N$.

Now consider that (7.15) gives us

$$\Sigma^{-1} = \begin{pmatrix} 1/\sigma_1^2 & & 0 \\ & \ddots & \\ 0 & & 1/\sigma_N^2 \end{pmatrix} \qquad \text{and} \qquad |\Sigma| = \prod_{j=1}^{N} \sigma_j^2.$$

The first of these enables us to write the quadratic form in the exponent of the Multivariate Normal density function of $\mathbf{X}$ as

$$(\mathbf{x} - \boldsymbol{\mu})^\top \Sigma^{-1} (\mathbf{x} - \boldsymbol{\mu}) = \sum_{j=1}^{N} (x_j - \mu_j)^2 / \sigma_j^2$$

and the second to write the coefficient of the exponential as

$$\left[(2\pi)^N |\Sigma| \right]^{-1/2} = \left[(2\pi)^N \prod_{j=1}^{N} \sigma_j^2 \right]^{-1/2} = \left[\prod_{\sigma=1}^{N} 2\pi \sigma_j^2 \right]^{-1/2}$$

so that

$$f_{\mathbf{X}}(\mathbf{x}) = \left[\prod_{j=1}^{N} 2\pi \sigma_j^2 \right]^{-1/2} e^{-(1/2)\sum_{j=1}^{N}(x_j - \mu_j)/\sigma_j^2} = \prod_{j=1}^{N} \frac{1}{\sqrt{2\pi \sigma_j^2}} e^{-(x_j - \mu_j)^2 / 2\sigma_j^2}.$$

The latter expression is the product of the marginal distributions of the components $\mathbf{X}$, as claimed.   ∎

It follows, by the definition of independence of continuous variables given in Chapter 3, that $X_1, \ldots, X_N$ are independent.

We have established that for variables with a joint Multivariate Normal distribution, diagonality of the variance/covariance matrix implies stochastic independence of the variables. It follows that if a linear transformation of $\mathbf{X} \to \mathcal{N}(\boldsymbol{\mu}, \mathbf{\natural})$ diagonalizes $\mathbf{\natural}$,

$$Q\mathbf{\natural}Q^{\top} = \begin{pmatrix} \tau_1^2 & & 0 \\ & \ddots & \\ 0 & & \tau_N^2 \end{pmatrix}$$

then the transformed variable $\mathbf{Y} = \mathbf{QX}$ has independent components. That is, if $\mathbf{q}_j^{\top}$ and $\mathbf{q}_k^{\top}$ are row vectors of $\mathbf{Q}$, then $\mathbf{q}_j^{\top}\mathbf{X}$ and $\mathbf{q}_k^{\top}\mathbf{X}$ are stochastically independent scalar variables for $k \neq j$. In addition we know that each $\mathbf{q}_j^{\top}\mathbf{X}$ has Univariate Normal distribution with mean $\mathbf{q}_j^{\top}\mu_j$ and variance τ_j^2.

Distributions of Differences between Means of Independent Data Sets

The distribution of the difference between the means of values observed in different circumstances can be obtained directly as a consequence of (7.15). The notation we use here may be a bit confusing at first. However, let us designate the observations from one set of circumstances as $X_1, \ldots, X_M$ and the observations from another as $Y_1, \ldots, Y_N$. These might be surface air temperatures measured from two ships: one to the northwest of the Gulf Stream and the other to the southeast. We may suppose that the measurements were made at approximately the same times of day, throughout the period of an experiment. Nonetheless we know that they were obtained from different instruments, by different people, on ships with differently configured work areas and schedules. If we assume that the vector of combined observations, $(X_1 \ldots X_M Y_1 \ldots Y_N)$, has a Multivariate Normal distribution for which the X's are all uncorrelated with the Y's, we may use the transformation of this vector that gives us $\bar{X} - \bar{Y}$, to obtain the distribution of the difference between the means. Specifically,

$$\bar{X} - \bar{Y} = Q \begin{pmatrix} X_1 \\ \vdots \\ X_M \\ Y_1 \\ \vdots \\ Y_N \end{pmatrix} \quad \text{for } Q = \left(\frac{1}{M} \cdots \frac{1}{M} \frac{-1}{N} \cdots \frac{-1}{N} \right). \quad (7.16)$$

We will suppose that the means of the X's may be different for each of the days and locations of the cruise of the ship to the NW of the Gulf Stream, and denote them by $\mu_1, \ldots, \mu_M$; and similarly for the Y's from the cruise of the ship to the SE of the Gulf Stream, for which we denote the means by $\nu_1, \ldots, \nu_N$. The correlations between pairs of values will be denoted by ρ_{jk} for X_j and X_k, and by ζ_{jk} for Y_j and Y_k. Finally, we assume that the increments for the X's, $(X_j - \mu_j)$, have common variance value σ^2 and the increments for the Y's, $(Y_j - \nu_j)$, have common variance τ^2, which may very well be different from σ^2. Now we can write

$$\mathcal{E}\left[(X_j - \mu_j)(X_k - \mu_k)\right] = \rho_{jk}\sigma^2$$

and

$$\mathcal{E}\left[(Y_j - \nu_j)(Y_k - \nu_k)\right] = \zeta_{jk}\tau^2.$$

Putting our assumptions and their consequences together in a single statement about the joint distribution, we have

$$\begin{pmatrix} X_1 \\ \vdots \\ X_M \\ Y_1 \\ \vdots \\ Y_N \end{pmatrix} \to \mathcal{N}\left(\begin{pmatrix} \mu_1 \\ \vdots \\ \mu_M \\ \nu_1 \\ \vdots \\ \nu_N \end{pmatrix}, \begin{pmatrix} \sigma^2 \; \sigma^2\rho_{12} \cdots \sigma^2\rho_{1M} & 0 \\ \ddots & \\ \sigma^2 & \\ & \tau^2 \tau^2\zeta_{12} \cdots \tau^2\zeta_{1N} \\ 0 & \ddots \\ & \tau^2 \end{pmatrix} \right).$$

Now we apply (7.15), with the Q of (7.16), and obtain

$$\bar{X} - \bar{Y} = Q\begin{pmatrix} \mathbf{X} \\ \mathbf{Y} \end{pmatrix} \to \mathcal{N}\left(Q\begin{pmatrix} \boldsymbol{\mu} \\ \boldsymbol{\nu} \end{pmatrix}, Q\Sigma Q^\top \right),$$

where

$$Q\begin{pmatrix} \boldsymbol{\mu} \\ \boldsymbol{\nu} \end{pmatrix} = \sum_{j=1}^{M} \mu_j \Big/ M - \sum_{j=1}^{N} \nu_j \Big/ N$$

and

$$Q\Sigma Q^\top = \frac{\sigma^2}{M^2}\sum_{j,k}\rho_{jk} + \frac{\tau^2}{N^2}\sum_{j,k}\zeta_{jk}. \tag{7.17}$$

In those special circumstances in which it is appropriate to assume that for all $j \neq k$

$$\rho_{jk} = 0 = \zeta_{jk}$$

(7.17) simplifies to

$$\bar{X} - \bar{Y} \to \mathcal{N}(\bar{\mu} - \bar{\nu}, \; \sigma^2/M + \tau^2/N). \tag{7.18}$$

Regarding Normal Distributions

In concluding this long chapter it may be helpful to reflect on the material presented here. It concerns the foundation of much of the estimation and hypothesis testing of ocean and atmospheric sciences. We have reviewed what we mean when we say that a variable is normally distributed and surveyed tests which can help to determine whether this is an accurate description for a particular data set. We have considered alternative formulations for Multivariate Normal distributions; and we have explored salient properties and discovered keys to data analysis in the composition of the matrix which describes variabilities and covariabilities of the components of vector-valued variables. We exit the chapter knowing that there always exists a linear transformation to another normally distributed variable whose components are independent in the statistical sense. Although we will not pursue it in this text, the transformation to independent components is the basis of "principal component analysis," a topic superbly covered by Preisendorfer and Mobley (1988).

The following chapter introduces distributions which have been developed as practical inferential tools for use when complete specification of a Normal distribution is not available. For our purposes, this is the general rule.

EXERCISES

1. (a) From your own scientific background, present reasoning which would lead you to expect that the distribution of successive January average temperatures might be well approximated by a Normal distribution.

   (b) Use the 94YEARS file of monthly means to confirm your expectation. Create a two-column subfile, with the first 47 years of January means of daily average temperatures in the first column and the second 47 years in the second column. Use the DESCRIBE command as well as the HISTOGRAM command, so that you will have the values of the summary statistics in addition to the visual data displays. Would you say that your expectations are supported by the evidence you have generated here?

2. (a) Confirm by differentiating the general form of the characteristic function for a Normal distribution (7.2) that

$$\mathscr{E}[X] = \mu \quad \text{and} \quad \mathscr{E}[X^2] = \sigma^2 + \mu^2.$$

   (b) Using the mean and standard deviation values obtained in Exercise 1(b), for the Januarys in the 94YEARS file, write the characteristic functions for the January means of the first 47 years and the second 47 years.

   (c) Again with the means and standard deviations for the two sets of Januarys from 94YEARS, use MINITAB to generate values of the corresponding p.d.f.s.

Hint: Create a MINITAB file with the "SET C1" command, putting the values 26.0, 28.0, 30.0, 32.0, 34.0, 36.0, 38.0, 40.0, 42.0, 44.0, 46.0, 48.0 into the first column, not forgetting "END" at the end of the data. These will be the values at which the p.d.f.s are calculated. Then use the "PDF C1 C2; NORMAL $\bar{X}_1, S_1$." command; and repeat it as "PDF C1 C3; NORMAL $\bar{X}_2, S_2$.", not forgetting the periods at the ends of the parameter specifications. These two commands put the density function values into columns C2, for the first 47 years of January mean temperatures, and C3 for the second 47 years. You can now print the three columns C1, C2, and C3, and plot the curves C2 vs C1 and C3 vs C1.
Further hint: Handmade plots on graph paper will provide better illustrations than MINITAB's PLOT command, because the p.d.f.s are smooth curves and MINITAB cannot draw smooth curves.

3. With State College data from the TEMPS file for the single January of 1986, create a histogram, a set of descriptive statistics, and a corresponding plot of the Normal distribution curve with matching mean and standard deviation. Look at them carefully. If you would feel confident using the Normal distribution to represent the distribution of this data, present supporting arguments based on your results. If you would not, explain your reasoning.
Hints for the constructions: Read the directions in Exercise 2(c).

4. (a) For a monthly mean temperature variable with $\mu = 5$ and $\sigma^2 = 9$, evaluate the following probabilities by converting to $Z = (X - \mu)/\sigma$ and using the $N(0, 1)$ table: (i) $\Pr[X \leq x]$ for $x = -1, 2, 5, 8, 11$; (ii)$\Pr[2 < X \leq 8]$ and $\Pr[-1 < X \leq 11]$.
   (b) Evaluate the probabilities of (a)(i) with MINITAB. Set the values into C1 and use the CDF C1 C2;... command. Write the command on your homework paper and copy over the values of columns 1 and 2. Compare with your results for (a).

5. (a) Put the values $-3, -2, -1, 0, 1, 2, 3, 4, 5, 6, 7, 8, 9, 10, 11, 12, 13$ into C4. With $\mu = 5.714$ and $\sigma = 3.268$, evaluate the c.d.f. at each of the values in C4 and put them into C5. Label and print these two columns, outline in red, and place the results in your homework paper.
   (b) With the mean and standard deviations of part (a), find the "p-values" for $p = 0.5, 0.75, 0.90, 0.95, 0.99$, using "INVCDF...". Interpret the results in a table, as

x	$\Pr[X \leq x]$	$\Pr[X > x]$
$\vdots$	$\vdots$	$\vdots$

6. (a) Use STN#27 salinity data at the surface and at 125 m, separately: 12 values at each level. Determine whether it is reasonable to treat these data sets as having normal distributions. Give both visual and quantitative support for your conclusions.

   (b) Repeat part (a) for the temperature values at the same two levels. If your conclusion is that one or both of the data sets cannot be treated as having a normal distribution, can you find a transformation of the data which passes your visual and quantitative criteria for normality?

   (c) Review the conditions under which we can expect the distribution constructed from an observed set of data to be well approximated by a Normal distribution. Is it reasonable to think these conditions are met for the STN#27 data? Explain. If there is any reason you would not expect these data to meet the criteria, present your argument.

7. Use daily temperature ranges from BOISE (created from CLIMAT in Exercise 2.5).

   (a) Create a separate MINITAB file having the day of the year and the temperature range values for the first 60 days of 1990, in C1 and C2. Plot C2 versus C1; and make a histogram and "describe" the data in C2.

   (b) For these 60-days' temperature range values, determine whether or not it is reasonable to treat them as having a Normal distribution. What information obtained with the DESCRIBE command supports your conclusion?

   (c) Use just the data for February. Obtain the maximum difference between the observed distribution and the normal distribution matched to the observed February mean and variance. Compare this with the values in Table 7.3, and say whether you would feel confident in assuming the distribution of February 1990 temperature ranges is a "sample" from a Normal distribution. If there is anything about this assumption that is "counter intuitive," explain. (You are probably right.)

8. Use daily maximum and minimum temperature values for Little Rock (LIT) from the TEMPS file.

   (a) Choose the April values for 1978–1987. Compute the means of MAXT and MINT, for each of the 10 years, and create a new MINITAB file with the years in C1 and the mean values in C2 and C3. Use the DESCRIBE and CORRELATE commands to evaluate parameters of the joint distribution of April mean daily maximum and minimum temperatures. Write these in vector and matrix array form, assuming the joint distribution is Bivariate Normal.

(b) Write the formulation for the Bivariate Normal density function, using the parameter values determined in part (a).

(c) Write the expression for the characteristic function for the corresponding Bivariate Normal distribution.

9. From the CLIMAT file, choose Spring in NOME and select the values for MAXT, MINT, and PKWSP, to create a new MINITAB file.

(a) Use the DESCRIBE and CORRELATE commands to find the means, variances, and correlations of these three variables; and write them in vector and matrix arrays.

(b) Assuming that the three variables have a joint Normal distribution, write out the expression for the density function, using the parameter values you obtained in part (a).

(c) Write the expression for the corresponding characteristic function.

10. From the CLIMAT file, choose Spring in NEW (Orleans) and select the values for AVT, TD, and AVWSP, to create a new MINITAB file.

(a) Use the DESCRIBE and CORRELATE commands to find the means, variances, and correlations of these three variables; and write them in vector and matrix arrays.

(b) Assuming that the three variables have a joint Normal distribution, write out the expression for the density function, using the parameter values you obtained in part (a).

(c) From an examination of the parameters, would you say that any of the pairs of variables here may be independent, in the stochastic independence sense. If so, which ones? What factorization of the joint density function does this permit?

11. (a) Confirm that $\Sigma = BB^{\top}$ for

$$
\Sigma = \begin{pmatrix} 4 & 2 & 2 \\ 2 & 10 & 1 \\ 2 & 1 & 2 \end{pmatrix} \quad \text{and} \quad B = \begin{pmatrix} 2 & 0 & 0 \\ 1 & 3 & 0 \\ 1 & 0 & 1 \end{pmatrix}.
$$

(b) If

$$
\begin{pmatrix} X_1 \\ X_2 \\ X_3 \end{pmatrix} \to \mathcal{N}\left(\begin{pmatrix} 1 \\ 0 \\ 1 \end{pmatrix}, \Sigma \right)
$$

find a transformation of X that has mean

$$
\begin{pmatrix} 0 \\ 0 \\ 0 \end{pmatrix}
$$

and variance/covariance matrix

$$I = \begin{pmatrix} 1 & 0 & 0 \\ 0 & 1 & 0 \\ 0 & 0 & 1 \end{pmatrix}.$$

Write the elements of the transformed variable **Z** as linear combinations of the X_j's.

(c) What is the characteristic function for **Z**?

12. (a) Using $\boldsymbol{\ddagger}$ from Exercise 11, find the mean and variance of $\bar{X}$.
    (b) What is the characteristic function for $\bar{X}$?
    (c) What is the characteristic function for $Y = 3(\bar{X} - 2)$; and what are the mean and variance of Y?

13. Use the data set 94YEARS.
    (a) Create a set of January mean temperatures, with one value for each year. Make two subsets with 40 values in each: one subset for the years 1910–1949 and one subset for the years 1950–1989. (Name these EARLIER and LATER, and save them. Discard the remaining mean values.) Create and label histograms for these two subsets, and place them in your homework paper.
    (b) (i) Did you anticipate that the histograms you created would look like sample distributions of normally distributed variables? Why? (ii) Are you satisfied that they do appear to have come from a Normal distribution? Explain.
    (c) Think of the data as $X_1, \ldots, X_{40}$, for 1910–1949, and $Y_1, \ldots, Y_{40}$, for 1950–1989. Estimate the means and variances of the distributions of $\bar{X}$ and $\bar{Y}$, assuming that they remained constant throughout each 40-year period. Identify the distribution of $(\bar{X} - \bar{Y})$, including its mean and variance.
    (d) Write a "diary note" describing the steps by which you reached the end result. Include both reasoning and computational steps.
    (e)–(g) Repeat (a)–(c) for the July values in 94YEARS.

14. Use salinity values of the data set STN#27.
    (a) For each of the 12 months, the 10 values of salinity give us a (10×1) vector whose elements are correlated and have different means and variances. If we assume that the correlations (with depth), and the means and variances which are specific to level, are constant throughout the 12 months, these can be estimated for the (10×1) vectors. Do this and create the (10×10) matrix $\boldsymbol{\ddagger}$.
    (b) Now describe the distribution of the difference between the annual means for salinity of the top and bottom depths.

15. Use AVWDR and PKWSP values for the days of Spring from the CLIMAT data set. Create two columns of 91 values, for each of the four stations Nome, Manistee, New Orleans, and Charlotte; and regard these as providing 91 observations of the vector

$$
\left(
\begin{array}{l}
\left(\begin{array}{l} \text{AVWDR} \\ \text{PKWSP} \end{array}\right)_{\text{Nome}} \\
\left(\begin{array}{l} \text{AVWDR} \\ \text{PKWSP} \end{array}\right)_{\text{Manistee}} \\
\left(\begin{array}{l} \text{AVWDR} \\ \text{PKWSP} \end{array}\right)_{\text{New Orleans}} \\
\left(\begin{array}{l} \text{AVWDR} \\ \text{PKWSP} \end{array}\right)_{\text{Charlotte}}
\end{array}
\right).
$$

Use the values in these columns to estimate the corresponding mean vector $\boldsymbol{\mu}$ and the matrix Σ. You may assume that the daily wind values at different stations are stochastically independent. Write out the characteristic function for the distribution of this (8×1) vector of wind values, assuming that the distribution is Multivariate Normal.

Hint: The product of four Bivariate Normal distributions will be the easiest to manage. Why is this justified?

8

ANALYZING VARIABILITY
"establishing differences between means and between variances"

8.1 INTRODUCTION

Continuing from the last chapter, we now take up the challenge of making inferences about ensemble properties of variables for which we have collections of observations. Use of the techniques we introduce here necessarily assumes that *the distributions describing our observed variables are Normal distributions*. While that may be considered an undesirably restrictive assumption in some cases, it is the foundation from which the distributions of the χ^2, t, and F statistics have been derived. This chapter provides an introduction to the defining conditions and descriptions of the namesake distributions of the χ^2, t, and F statistics, and their use in the construction of confidence intervals for variance and mean values, differences between means, and ratios of variances. This is powerful equipment to support our quest for quantifying differences between comparable systems observed at different times or locations, with assignable measures of confidence in ascribed distinctions.

Inference and Confidence Intervals

A *confidence interval*, whether it is for a variance parameter, σ^2, a difference between means, $\mu - \nu$, or a ratio of variances, σ^2/η^2, *is a set of values determined by the data set as most likely for the parameter in question*. Associated with the interval is the probability P that it contains the true value of the parameter. Since the numbers in the data set are observed values of a stochastic variable and the interval is denoted by end point statistics which are calculated from the data, the interval itself is a stochastic variable. A different set of observations will give us a different interval and, because of the influence of random variability, not every possible interval will include the true value of the parameter for which it is constructed. *Given the values observed for the variable under study, P is our measure of confidence that the interval constructed from the observations actually contains the parameter in question*. The value designated for P is our choice. Generally we use a relatively large value of P, since we wish to be quite confident that the interval has "captured" the true value of the parameter. The interval then designates those values which we consider credible values for the parameter, based on the data used to construct it.[1]

Examples of Inference Objectives

A few examples, for reference, may help in tying down the material of this chapter. Consider the following as possible objectives.

(i) Determination of whether continental temperatures in the decades since 1949 exhibit greater variability than that of the first half of the 20th century. This will be achieved by comparison of the variance of observed temperatures for 1950 onwards, with the value of the variance established for 1900 to 1949.

(ii) Quantification of vertical salinity differentials in the northwest Atlantic, by putting bounds on mean layer salinity differences.

(iii) Comparison of observed variances for recorded wind directions at potential wind power generation sites, through knowledge of the distributions of ratios of observed variances.

We will return to these examples in illustrations and exercises, and address them with data from our common files.

Inferential Statistics from a "Normal Foundation"

The Multivariate Normal distribution has been presented in some detail, in the foregoing chapter, because it is the theoretical basis for the

---

[1]If you are saying to yourself, "That sounds rather like hypothesis testing," you are right on track. The rationale for construction of confidence intervals is closely allied with principles of testing hypotheses. The direct relationship between them will be established in Chapter 9.

distributions of χ^2, t, and F statistics. It provides the framework within which we find and validate the distributions of statistics that serve our present inference objectives. We may characterize the three distributions of this chapter, in words, as follows.

1. *The Chi Square* is the distribution of a scaled ratio of the variance of a set of observations and the true variance of the observed variable.
2. *Student's t* is the distribution of a standardized difference between the mean of a set of observations and the true mean of the observed variable. The t statistic and its distribution apply as well to differences between means for different data sets, following the example at the end of Chapter 7.
3. *Fisher's F* is the distribution of the ratio of scaled variances of independent data sets and of independent estimates of the same variance, each divided by the degrees of freedom of the estimate. It provides the key for establishing differences between the values of true variances, as well as the relative impacts of different sources of variability.

8.2 THE QUADRATIC FORM: KEYSTONE IN ANALYSES OF VARIABILITY

In the order given, χ^2, t, and F statistics and their distributions will be defined and then used to construct confidence intervals for values of the following parameters:

(i) a variance,
(ii) a mean or a difference between means,
(iii) a ratio of variances.

This is done in the context of an analysis of the variability within a composite data set. Since the variability within a data set is a direct reflection of the many (and sometimes disparate) influences that shape observed phenomena, the tools we develop here may be used to attribute portions of the differences among observations to identifiable sources—with associated measures of confidence.

To begin, we define *the quadratic form* (QF) *for a Multivariate Normal vector* $\mathbf{X} \in \mathscr{R}^N$, as the redoubtable expression

$$(\mathbf{X} - \boldsymbol{\mu})^T \, \boldsymbol{\Sigma}^{-1} (\mathbf{X} - \boldsymbol{\mu}). \tag{8.1}$$

Whatever the values of its elements are, in a specific situation, this is a scalar-valued measure of total variability for the increment between the (vector-valued) observed variable and its (vector-valued) ensemble mean.

As such, the quadratic form is the cumulative measure which we wish to partition—into components associated with accountable sources of variability, plus a residual of unaccountable variability.

At this point of discussion, we will not be concerned with the mechanics of partitionings. However, as an aid to thinking about inference objectives in relation to the quadratic form, recall that we have already suggested that there may be an influence of chronological time period on continental temperature variability. The evaluation (statistical test) of this possibility will use a partitioning of the data set. This is equivalent to a partitioning of the cumulative quadratic form for the whole system, into components associated with comparative time periods.

By using $\Lambda_{N \times N}$ to denote the inverse of the matrix of variances and covariances

$$(\lambda_{jk}) = \Lambda = \Sigma^{-1}$$

we may rewrite (8.1) as

$$(\mathbf{X} - \boldsymbol{\mu})^T \Lambda (\mathbf{X} - \boldsymbol{\mu}) = \sum_{j,k} (X_j - \mu_j) \lambda_{jk} (X_k - \mu_k). \qquad (8.2)$$

In the form given by (8.2) we can think of grouping terms of the sum together into factors associated with different sources of variability. In fact, each of the three statistics on which we focus in this chapter is obtained as a combination of factors of the quadratic form.

Each statistic will have its namesake distribution under the conditions for the validity of

$$\mathbf{X} \to \mathcal{N}\left(\boldsymbol{\mu}, \Sigma \right). \qquad (8.3)$$

Note that the QF is defined by the replacement of $\mathbf{x}$ by $\mathbf{X}$ in the term

$$(\mathbf{x} - \boldsymbol{\mu})^T \Sigma^{-1} (\mathbf{x} - \boldsymbol{\mu})$$

which is in the exponent of the Normal probability density function. Also note that it contains in its definition all the parameters of the assumed distribution of $\mathbf{X}$.

When (8.3) is valid, the distribution which describes the stochastic properties of the quadratic form in its entirety is the Chi Square distribution with parameter equal to the number of degrees of freedom. We write

$$\chi_N^2 = (\mathbf{X} - \boldsymbol{\mu})^T \Sigma^{-1} (\mathbf{X} - \boldsymbol{\mu}). \qquad (8.4)$$

All the parameters associated with the variable $\mathbf{X}$ are included in the expression for χ^2; and its distribution depends only on the dimension of the space of possible values for the vector $\mathbf{X}$. This is also the rank of the matrix $\ddagger$. Anticipating its use in the future, we introduce a more general notation for a Chi Square distributed statistic, χ_κ^2, where κ is the number of identifiable, independent sources of variability.

The mathematical expressions that describe the distribution of χ_κ^2 in terms of its probability density and characteristic function may be found elsewhere. (For example, see Kendall and Stuart, 1963, or *The Encyclopedia of Statistical Sciences*, 1982, which also give expressions of t and F distributions.) Since we will not use these functions explicitly, we will not copy them here. Statistical software packages and published statistical tables, such as the tables edited by Pearson and Hartley (1966) and Fisher and Yates (1967), give us the access to these distributions that we require for their use. Plots of their density functions are shown below, in the sections which explore the uses of χ^2, t, and F.

Partitioning the Quadratic Form

Let's assume that the components of **X** comprise a set of values $X_1, \ldots, X_N$, observed or to be observed during the course of a field experiment. Notice that if we could apply the linear transformation with matrix **A** from the factorization

$$\ddagger = \mathbf{BB}^T \qquad \text{with } \mathbf{A} = \mathbf{B}^{-1} \tag{8.5}$$

to define

$$\mathbf{Y} = \mathbf{AX} \qquad \text{and} \qquad \boldsymbol{\nu} = \mathbf{A}\boldsymbol{\mu}$$

then

$$(\mathbf{X} - \boldsymbol{\mu})^T \ddagger (\mathbf{X} - \boldsymbol{\mu}) = [\mathbf{A}(\mathbf{X} - \boldsymbol{\mu})]^T [\mathbf{A}(\mathbf{X} - \boldsymbol{\mu})]$$

$$= (\mathbf{Y} - \boldsymbol{\nu})^T (\mathbf{Y} - \boldsymbol{\nu}) = \sum_{j=1}^{N} (Y_j - \nu_j)^2.$$

The last expression measures the total variation of the Y_j's around their own expected values. Because $\ddagger$, and therefore **B** and **A** as well, are nonsingular, the transformation to **Y** does not reduce the degrees of freedom. Hence this alternate representation supports interpretation of the QF as a composite expression of the total, accountable variation of the observed system. Of course, in order to make the transformation that produces this representation, we would require full knowledge of $\ddagger$, which we generally do not have. Nonetheless we can approach the objective of analyzing the total variation with a mathematical partitioning of the QF into components associated with identifiable sources of variability. Then we construct statistics with which we can measure the relative impacts of these sources.

Cochran's Theorem, which is arguably the most important theorem of statistical data analysis, provides the cornerstone for analyses that associate components of total observed variability with identifiable sources.

ASSERTION 8.1 (COCHRAN'S THEOREM). *For sources of variability indexed by* k, *suppose that we can write*

$$(\mathbf{X} - \boldsymbol{\mu})^T \boldsymbol{\Sigma}^{-1} (\mathbf{X} - \boldsymbol{\mu}) = (\mathbf{Y} - \boldsymbol{\nu})^T (\mathbf{Y} - \boldsymbol{\nu}) = W_1^2 + \cdots + W_K^2,$$

with a reduced number of components on the right, i.e., with $K < N$. *Let* N_k *be the rank of* W_k, *for each* $k = 1, \ldots, K$. *Then* $W_1^2, \ldots, W_K^2$ *will have independent Chi Square distributions*

$$W_k^2 = \chi_{N_k}^2$$

if, and only if, $N_1 + \cdots + N_K = N$.

A proof of this extremely powerful result is given by Scheffé (1959, pp. 420–421). In words, the theorem says that *if we can partition the QF or, equivalently, the total sum of the squared deviations of the Y's from their mean values into a sum of quadratic forms for which the degrees of freedom of the summands add to the original degrees of freedom, then the component QFs are independent Chi Square variables.*

As an example of the utility of Cochran's Theorem, suppose that

$$\mathbf{Y} \to \mathcal{N}\left(\nu \begin{pmatrix} 1 \\ \vdots \\ 1 \end{pmatrix}, \sigma^2 \begin{pmatrix} 1 & & 0 \\ & \ddots & \\ 0 & & 1 \end{pmatrix}\right). \tag{8.6}$$

The scalar components of $\mathbf{Y}$ have common mean ν and variance σ^2, and

$$\sum_{j=1}^{N} (Y_j - \nu)^2 / \sigma^2 = \chi_N^2.$$

With

$$\bar{Y} = \sum_{j=1}^{N} Y_j / N$$

we may rewrite and factor the sum of squares as

$$= \sum_j \left[(Y_j - \bar{Y}) + (\bar{Y} - \nu) \right]^2 / \sigma^2$$

$$= \left[\sum_j (Y_j - \bar{Y})^2 + 2(\bar{Y} - \nu) \sum_j (Y_j - \bar{Y}) + N(\bar{Y} - \nu)^2 \right]$$

$$= \sum_j (Y_j - \bar{Y})^2 / \sigma^2 + 0 + (\bar{Y} - \nu)^2 / (\sigma^2 / N)$$

$$= W^2 + Z^2.$$

You will recognize W^2 as $(N-1)S^2/\sigma^2$ and Z^2 as the square of the standardized mean of the observations

$$Z = (\bar{Y} - \nu)\bigg/\sqrt{\frac{\sigma^2}{N}}\,.$$

Cochran's Theorem applies here, with $(N-1)$ degrees of freedom for W^2 and 1 degree of freedom for Z^2. Thus we have proved

ASSERTION 8.1.

$$(N-1)S^2/\sigma^2 = \chi^2_{(N-1)} \quad \text{and} \quad (\bar{Y} - \nu)^2\bigg/\left(\frac{\sigma^2}{N}\right) = \chi^2_1;$$

and these functions of the variance and mean of the observations are stochastically independent statistics.

The foregoing is an extremely important result. However, it may also seem a bit surprising, since the computation of both statistics uses the full set of observations.

In the sections that follow we will focus on the Chi Square distribution of the ratio of the variance of a set of observations to the true variance of the observed variable, S^2/σ^2; on Student's t distribution of a standardized mean of a set of observations, with the true variance estimated by S^2,

$$(\bar{X} - \mu)/\sqrt{S^2/(N-1)} = \sqrt{N-1}\,(\bar{X} - \mu)/S;$$

and on Fisher's F distribution of the ratio of independent estimates of the same variance. Each is derived from factors in a partitioning of the total variation QF.

8.3 THE CHI SQUARE DISTRIBUTION AND CONFIDENCE INTERVALS FOR A VARIANCE PARAMETER

The Chi Square is probably best known as the distribution of

$$\sum_{j=1}^{N}\left(Y_j - \bar{Y}\right)^2/\sigma^2 = (N-1)S^2/\sigma^2 = \chi^2_{(N-1)}$$

as it is used to make inferences about the value of σ^2, when (8.6) describes the stochastic behavior of **Y**. In words, the assumptions are that the scalar components of **Y** are stochastically independent variables, with common mean and common variance values. Under these conditions we may obtain probabilities for any interval of possible values for $\chi^2_{(N-1)}$. We note that by its definition it is a positive-valued statistic. Examples of its density functions for various values of the parameter $\kappa = (N-1)$ are shown in Fig. 8.1.

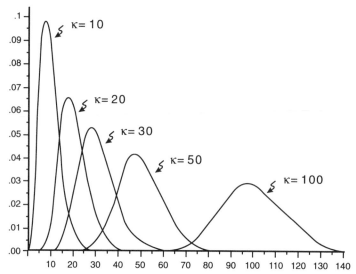

FIGURE 8.1 Density functions for the Chi Square distribution, with degrees of freedom $\kappa = 10, 20, 30, 50, 100$.

Let's suppose that $\kappa > 1$ (which it generally will be since we can't do much with just 1 degree of freedom) and that we wish to identify the endpoints of an interval "central" to the distribution of χ^2. Since the distribution is asymmetric we must first decide what "central" is going to mean for us, in terms of apportioning probability to the left and right of the interval endpoints. For example, an interval for χ^2 that contains 80% of the probability, leaves 20% to be divided between the left and right ends of the distribution. One option, although not the only reasonable one, is to apportion it equally. For the equal allocation case, we find the value preceded by 10% of the probability and the value preceded by 90% of the probability (leaving 10% to the right of the right endpoint). We can obtain these values with MINITAB commands, using the appropriate value of κ:

"INVCDF 0.10: CHISQUARE κ." and
"INVCDF 0.90: CHISQUARE κ."

If we denote MINITAB's responses by $\chi^2_{\kappa;0.10}$ and $\chi^2_{\kappa;0.90}$, respectively, then what we have found are the values for which

$$0.10 = \Pr\left[\chi^2_\kappa \leq \chi^2_{\kappa;0.10}\right] \quad \text{and} \quad 0.90 = \Pr\left[\chi^2_\kappa \leq \chi^2_{\kappa;0.90}\right].$$

We may combine them to write

$$\Pr\left[\chi^2_{\kappa;0.10} < \chi^2_\kappa \leq \chi^2_{\kappa;0.90}\right] = \Pr\left[\chi^2_\kappa \leq \chi^2_{\kappa;0.90}\right] - \Pr\left[\chi^2_\kappa \leq \chi^2_{\kappa;0.10}\right] = 0.80.$$

Thus $\chi^2_{\kappa;0.10}$ and $\chi^2_{\kappa;0.90}$ are the endpoints of a central interval containing 80% of the distribution of χ^2_{κ}.

To convert the foregoing to a confidence interval for the unknown value of the variance, we note that by substituting $(N-1)$ for κ and $(N-1)S^2/\sigma^2$ for χ^2_{κ}, we have

$$0.80 = \Pr\left[\chi^2_{(N-1);0.10} < \frac{(N-1)S^2}{\sigma^2} \le \chi^2_{(N-1);0.90}\right].$$

Now taking reciprocals of the terms within the square brackets [ ], which will reverse the directions of the inequalities, and multiplying through by $(N-1)S^2$, we arrive at our objective

$$0.80 = \Pr\left[(N-1)S^2/\chi^2_{(N-1);0.90} \le \sigma^2 < (N-1)S^2/\chi^2_{(N-1);0.10}\right]. \quad (8.7)$$

And we can evaluate the interval endpoints from the information in the data set.[2]

VIP: What we have done here has been to use the probability distribution of

$$\chi^2_{(N-1)} = (N-1)S^2/\sigma^2$$

to construct an interval for which we can say that *the probability is 0.80 that*

$$\left[(N-1)S^2/\chi^2_{(N-1);0.90},(N-1)S^2/\chi^2_{(N-1);0.10}\right) \quad (8.8)$$

contains the true value of the variance parameter σ^2. We call this *an 80% confidence interval for* σ^2.

If we chose to allocate end-region probabilities of 0.05 on the left and 0.15 on the right, the procedure would be to call MINITAB with

"INVCDF 0.05: CHISQUARE (N − 1)." and
"INVCDF 0.85: CHISQUARE (N − 1)."

and construct the 80% confidence interval

$$\left[(N-1)S^2/\chi^2_{(N-1);0.85},(N-1)S^2/\chi^2_{(N-1);0.05}\right).$$

We can illustrate this with an example derived from the 94YEAR data file for State College. In Exercise 1 of Chapter 7, descriptive statistics were

---

[2]When we substitute a value of S^2 computed from data, into this expression, the endpoints then have specific numerical values and it is no longer a probability statement in the strict sense. We call it a "fiducial probability statement".

obtained for January means of daily average temperatures for the two 47-year periods. For each of these we can construct 80% confidence intervals for the variances of the January means, that leave 5% of the probability in the left end of the χ^2 distribution and 15% in the right end. Since for both subsets of data $N = 47$, we use the commands

"INVCDF 0.05: CHISQUARE 46." and
"INVCDF 0.85: CHISQUARE 46."

to obtain the values $\chi^2_{46;\,0.05} = 31.44$ and $\chi^2_{46;\,0.85} = 55.92$ that we will use for both intervals. For the earlier period we found $S^2 = 23.33$; and for the later, $S^2 = 24.40$. Thus the two asymmetric 80% confidence intervals are

$$\left(\frac{46(23.33)}{55.92}, \frac{46(23.33)}{31.44} \right) = (19.19, 34.13), \qquad \text{for 1986 to 1942}$$

and

$$\left(\frac{46(24.40)}{55.92}, \frac{46(24.40)}{31.44} \right) = (20.07, 35.70), \qquad \text{for 1943 to 1989,}$$

intervals which have most of their values in common!

There are two special cases of particular importance for inferences about the variance parameter. One requires only an upper bound on the variance and the other requires only a lower bound, with associated probabilities that the one-sided intervals contain the true variance. These are generally used for comparison of the value calculated from a set of observations with a value known to characterize the distribution of the same variable in different circumstances or at a different time. We use one of these when our information is that the new variance is *either* smaller *or* larger, not just that we think it may have changed. (See Exercise 8.4.) Finding these "probability bounds" for a variance value is a bit tricky, since it involves taking reciprocals within the argument of the probability function. To show how it's done, we first apply the convention for designating the correspondence between values in the range of a statistic and its cumulative probabilities, to the Chi Square variable (8.7), in a general notation

$$p = \Pr\left[\chi^2_\kappa \leq \chi^2_{\kappa;\,p} \right]$$

and substitute

$$\frac{(N-1)S^2}{\sigma^2} \qquad \text{for } \chi^2_\kappa.$$

This gives us

$$p = \Pr\left[\frac{(N-1)S^2}{\sigma^2} \leq \chi^2_{(N-1); p}\right]$$

$$= \Pr\left[\sigma^2/(N-1)S^2 \geq 1/\chi^2_{(N-1); p}\right]$$

$$= \Pr\left[\sigma^2 \geq (N-1)S^2/\chi^2_{(N-1); p}\right] \tag{8.9}$$

and also, using the probability rule (4.16) of Chapter 4,

$$(1-p) = 1 - \Pr\left[\sigma^2 \geq (N-1)S^2/\chi^2_{(N-1); p}\right]$$

$$= \Pr\left[\sigma^2 < (N-1)S^2/\chi^2_{(N-1); p}\right]. \tag{8.10}$$

These expressions permit evaluation of specific probability bounds. For example, if we require a lower bound for the variance that puts 80% of the probability to the right of it, then we use (8.9) to get

$$0.80 = \Pr\left[\sigma^2 \geq (N-1)S^2/\chi^2_{(N-1); 0.80}\right],$$

where $\chi^2_{(N-1); 0.80}$ is output with the MINITAB command

"INVCDF 0.80: CHISQUARE (N − 1)."

If we need an upper bound for a variance interval with 80% of the probability to the left of it, then we must take $p = 0.2$ and use (8.10) to write

$$0.80 = \Pr\left[\sigma^2 < (N-1)S^2/\chi^2_{(N-1); 0.20}\right].$$

Again the χ^2 value is obtained from MINITAB, although this time with $p = 0.20$.

Note that we can rewrite the interval with both upper and lower bounds, in the general notation we used for one-sided intervals. For this we denote the "confidence value" as $(1 - p)$ and divide the probability outside the interval into equal increments of $(p\backslash 2)$. The endpoints of the $(1 - p) \times 100\%$ confidence interval[3] for σ^2 are determined by substituting $(N - 1)S^2/\sigma^2$ for $\chi^2_{(N-1)}$ and algebraically rearranging the argument of

---

[3]Established convention usually associates a percentage with the phrase "confidence interval", although probabilities are given as decimal numbers between 0 and 1. Thus, while we write $0.8 = \Pr[L(\mathbf{x}) \leq \sigma^2 < U(\mathbf{x})]$ to mean that the probability is 0.8 that the stochastic values $L(\mathbf{x})$ and $U(\mathbf{x})$ actually bracket the values of the parameter σ^2, we call $[L(\mathbf{x}), U(\mathbf{x}))$ "an 80% confidence interval for σ^2".

the probability function:

$$(1-p) = \Pr\left[\chi^2_{(N-1);\,p/2} < \chi^2_{(N-1)} \le \chi^2_{(N-1);\,(1-p/2)}\right]$$

$$= \Pr\left[\chi^2_{(N-1);\,p/2} < \frac{(N-1)S^2}{\sigma^2} \le \chi^2_{(N-1);\,(1-p/2)}\right]$$

$$= \Pr\left[(N-1)S^2/\chi^2_{(N-1);\,(1-p/2)} \le \sigma^2 < (N-1)S^2/\chi^2_{(N-1);\,p/2}\right].$$

$$(8.11)$$

Compare this expression with the two-sided 80% confidence interval we obtained above in (8.8), to see that this does indeed work.

An Example

We know from an archive of observations at State College, PA, going back to the nineteenth century, that December mean temperatures for the years preceding 1950 had a variance of 14. The data record for the next three decades, 1950 to 1979, suggest that there has been a significant increase in the variability of December mean temperature. That is, the observed variance computed for the December mean temperatures for this 30-year period is considerably larger, with a value of $S^2 = 18$. We wish to determine whether this larger value should be considered an artifact of the stochastic variability expected for December mean temperatures with true variance $\sigma^2 = 14$, or whether it constitutes evidence that the true variance is actually larger for the later period.

There are two courses open to us. We can be purists and set up a two-sided confidence interval for the true variance of the 1950–1979 period, as though we had not already examined the evidence, or we can use this evidence to select a one-sided comparison interval. In either case we must choose the level of confidence we wish to associate with the interval calculated for the 1950–1979 variance value. If the choice is 80% then, since $(N-1) = 29$ here, the MINITAB commands

"INVCDF 0.10: CHISQUARE 29." and
"INVCDF 0.90: CHISQUARE 29."

give us the two probability limits for the two-sided interval

$$\left[(N-1)S^2/\chi^2_{(N-1);\,0.90},\ (N-1)S^2/\chi^2_{(N-1);\,0.10}\right)$$

$$= \left[29S^2/39.88,\ \ 29S^2/19.77\right).$$

For the one-sided interval case, we use the MINITAB command with $p = 0.8$, so that we can write

$$0.80 = \Pr\left[\sigma^2 \ge (N-1)S^2/\chi^2_{(N-1);\,0.80}\right] = \Pr\left[\sigma^2 \ge 15\right].$$

Since the variance value from earlier years in the record is not in this 80% confidence interval, we may choose to conclude that the climate at State College has clearly altered in this respect.

8.4 STUDENT'S t DISTRIBUTION[4]: INFERENCES ABOUT MEANS AND DIFFERENCES BETWEEN MEANS

Student's *t* distribution is used to make inferences about the true mean of an observed variable when we do not assume prior knowledge about the variance. If we had reliable information about the true value of the variance, independent of the present data set, we would use the Normal distribution. However, generally we do not have such information; and we substitute variance values computed from observations in computation of the statistics used for inferences. Specifically, we formulate analogues to the standardized Normal variables

$$Z = (\bar{X} - \mu)/\sqrt{\sigma^2/M} \quad \text{and} \quad \left[(\bar{X} - \bar{Y}) - (\mu - \nu)\right]/\sqrt{\sigma^2/M + \tau^2/N},$$

by replacing σ^2 by S_X^2 and τ^2 by S_Y^2.

When the objective is to arrive at a judgement about the true mean of a single variable, with a set of observed values $X_1, \ldots, X_M$, we replace σ^2 by S_X^2 in the first of the standardized Normal statistics, to get

$$t = (\bar{X} - \mu)/\sqrt{S_X^2/M}. \tag{8.12}$$

This is *Student's t*. Its distribution has a single parameter: the degrees of freedom of the variance estimate.

The statistic *t* will have greater inherent variability than *Z*, because its formulation uses an estimate of the true variance, which has been calculated from the observed values of *X*. If there are not very many observations from which to estimate σ^2, the variability of *t* may be considerably greater than that of *Z*, because S^2 may differ from σ^2 by quite a bit. This is reflected in the breadth of the *t* distribution: A *t* statistic calculated from few observations has a broader distribution than a *t* statistic calculated from many observations. For large numbers of observations, the sample estimate of the variance will be dependably close to the true (but unknown) value of the variance; and the distribution of the *t* statistic will be correspondingly close to Normal, with $\sigma^2 = S^2$. In fact, when *M* is as large as 30, we ordinarily adopt the Normal distribution as a reasonable approximation to Student's *t*.

---

[4]"Student" was a pen name for W. L. Gossett who derived and published this family of distributions. The name Student has become a fixture in the statistical literature.

VIP: All of the distributions of Student's t are symmetric about zero, like the distribution of a standardized Normal variable.

A good way to remember the exact form of the statistic and to correctly identify the parameter is to remember that t is the ratio of $Z \to \mathcal{N}(0, 1)$ and $\sqrt{\chi^2_{(M-1)}/(M-1)}$:

$$t = \left[(\overline{X} - \mu)/\sqrt{\sigma^2/M} \right] / \sqrt{\left[(M-1)S^2/\sigma^2 \right]/(M-1)} .$$

The numerator is the standardized normal statistic for the sample mean $\overline{X}$. The denominator is the square root of the Chi Square variable $(M - 1)S^2/\sigma^2$ divided by its degrees of freedom.[5] The one parameter that we need to know to access the distribution of Student's t is the parameter of the χ^2 variable in the denominator. Here it is $(M - 1)$: the degrees of freedom for the variance estimate.

Construction of a Confidence Interval for μ

Because the distributions of t statistics are symmetric, it is generally reasonable to allocate equal probabilities to the two ends of a distribution in specifying a confidence interval. Again we will denote the measure of confidence as $(1 - p)$ and allocate $p/2$ to each end of the distribution, outside the confidence interval. By analogy with the two-sided intervals for variances obtained with the Chi Square distribution (8.11), we use Student's t distribution to bracket μ with probability $(1 - p)$. We substitute $(\overline{X} - \mu)/\sqrt{S^2/M}$ for t and rearrange the argument of the probability function, as shown, to obtain the desired interval:

$$1 - p = \Pr\left[t_{(M-1);p/2} < t \leq t_{(M-1);(1-p/2)} \right]$$

$$= \Pr\left[t_{(M-1);p/2} < \frac{\overline{X} - \mu}{\sqrt{S^2/M}} \leq t_{(M-1);(1-p/2)} \right]$$

$$= \Pr\left[\overline{X} - \sqrt{S^2/M}\, t_{(M-1);(1-p/2)} \leq \mu < \overline{X} + \sqrt{S^2/M}\, t_{(M-1);p/2} \right].$$

$$(8.13)$$

---

[5]As noted above, we must make the assumptions that the subject variables have Normal distributions *and* that the observations in each data set are stochastically independent of one another, with common mean and variance. These will not be reasonable assumptions in every case we wish to consider. However, it is important to acknowledge that they are the bases of the techniques introduced in this chapter. Procedures that provide alternative techniques for statistical inferences, when these assumptions are clearly violated, will be discussed in Chapter 11.

We picture this as

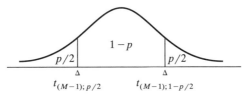

and note that the symmetry of the distribution around zero gives us

$$t_{(M-1); p/2} = -t_{(M-1); 1-p/2}.$$

Now, if we carry the degrees of freedom in our heads and denote the right-hand endpoint by $c_{p/2}$, the picture becomes

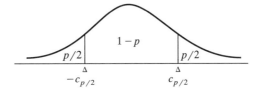

and we may write the confidence statement in the simpler notation

$$1 - p = \Pr\left[\overline{X} - c_{p/2}\sqrt{S^2/M} \le \mu < \overline{X} + c_{p/2}\sqrt{S^2/M} \right]. \quad (8.14)$$

The required values from Student's t distribution are obtained with the MINITAB command

$$\text{``INVCDF } \alpha; T\,k\text{.''} \quad (8.15)$$

by calling it with $\alpha = p/2$ and $k = M - 1$.

In general, with probability α allocated to each "tail" of Student's t distribution

$$\left[\overline{X} - c_\alpha\sqrt{S^2/M}, \overline{X} + c_\alpha\sqrt{S^2/M} \right)$$

is a $(1 - 2\alpha) \times 100\%$ confidence interval for the true mean μ. Thus, for a 85% confidence interval we would take $\alpha = .075$ and for a 95% confidence interval $\alpha = .025$.

An Example

Here we will use STN#27 data to bracket the true mean difference in salinities at 50 and 75 m, with an 85% confidence interval. From the salinity table, arranged by depth and month, we create a single column of data containing differentials $\Delta = s_{75} - s_{50}$. The assumptions that validate the use of the t distribution in the construction of the confidence interval apply to the differentials between 50 and 75 m, not to the salinity variables themselves. Specifically, we assume that the true mean and the true variance of the salinity differential are constant over the 12 months: the same for each of the $M = 12$ differences in observed averaged salinity values.

From the statistics returned by the MINITAB DESCRIBE command applied to the differential data, we get

$$\overline{\Delta} = 0.18 \quad \text{and} \quad S_{\Delta}^2 = 0.0104.$$

Now to compute the interval endpoints for the 85% confidence interval, with $M = 12$, we call

"INVCDF 0.075; T 11."

This returns $c_{0.075} = 1.58$, so that (8.14) gives us

$$0.85 = \Pr\left[\overline{\Delta} - 1.58\sqrt{S_{\Delta}^2/M} \leq \mu_{\Delta} < \overline{\Delta} + 1.58\sqrt{S_{\Delta}^2/M}\right] \quad (8.16)$$

and the 85% confidence interval for the true mean of the observed differential Δ is

$$\left[\overline{\Delta} - 1.58\sqrt{S_{\Delta}^2/M}, \overline{\Delta} + 1.58\sqrt{S_{\Delta}^2/M}\right) = [0.14, 0.23).$$

Alternate Computation

There is an alternative MINITAB command that will compute the end-points of the interval directly from the data when we supply both the column number in which the observations have been stored and the desired confidence level for the interval. MINITAB software will recompute the mean and variance of the observations, find the appropriate value of c_p, and carry out the assembly and arithmetic with the elements. The command is

"TINTERVAL PC C2"

for data in COLUMN 2 and percentage $PC = (1 - 2\alpha) \times 100\%$. In the example just given, if the observed values of the differentials have been copied into the second column of the current worksheet and we want an 85% confidence interval for the true mean difference between 75- and 50-m salinities, as we did above, then we call

"TINTERVAL 85 C2." (8.17)

Note that when this command is used, the confidence associated with the interval is expressed as a percent, rather than as a probability.

Construction of a Confidence Interval for a Difference between Means

Often we wish to compare characteristics of variables which have been observed in distinct locations or distinct circumstances, e.g., during different seasons or under markedly different physical conditions. To quantify the difference between true mean values when we have comparative data sets, $X_1, \ldots, X_M$ and $Y_1, \ldots, Y_N$, we would like to construct a t statistic from the standardized normal statistic

$$\left[(\bar{X} - \bar{Y}) - (\mu - \nu) \right] / \sqrt{\sigma^2/M + \tau^2/N}$$

substituting estimates of the variances: S_X^2 for σ^2 and S_Y^2 for τ^2. Certainly we can do this and form

$$t = \left[(\bar{X} - \bar{Y}) - (\mu - \nu) \right] / \sqrt{S_X^2/M + S_Y^2/N} \, .$$

However, there is a hazard in thinking that we know its distribution, unless we have been very careful with our diagnostics. If we are confident that the variances σ^2 and τ^2 are equal, then we can use Student's t distribution with $\kappa = (M - 1) + (N - 1)$ as the basis for constructing a confidence interval for the difference $(\mu - \nu)$. Consider, though, that since we are in a context in which we anticipate that the mean values are different, we may expect a difference in variances as well. If we use a test statistic for the difference in the means, which assumes equal variances, and the variances are not equal, then we have compromised our inference technology: *The test statistic does not have the distribution we have ascribed to it*. Generally we won't know what the distribution is; and we have lost the ability to make inferences with known measures of certainty. Thus it is important to use techniques especially developed for variables with (possibly) unequal variances.

If we cannot assume σ^2 and τ^2 are equal, then we have what is known as the Behren's–Fisher problem, for which there is not any totally satisfactory solution. Nonetheless, techniques for using a Student's t distribution as a reasonable approximation in circumstances in which it is unreasonable to assume $\sigma^2 = \tau^2$ have been developed. These techniques provide valuable alternatives to throwing one's hands up in despair. Brownlee (1965) presents and discusses a treatment which uses a function of the two sample variances to calculate a value for κ which gives

$$t = \left[(\bar{X} - \bar{Y}) - (\mu - \nu) \right] / \sqrt{S_X^2/M + S_Y^2/N}$$

a t_κ-like distribution. The formula for the calculated "degrees of freedom" parameter for this t statistic is

$$\kappa = \left[S_X^2/M + S_Y^2/N \right] / \left[(S_X^2/M)/(M - 1) + (S_Y^2/N)/(N - 1) \right].$$

However, you will probably never have to compute this number, unless you wish to be able to display it. Statistical software packages evaluate it and use it in determining points of the distribution, provided that you indicate that you wish an "unequal variances" statistical evaluation. The two-sample t algorithm in MINITAB automatically uses this formulation, with the rationale that it is better to err on the conservative side than to make inferences that are not supported by the data.

With the X and Y observations in columns C1 and C2 of a MINITAB worksheet, the command "TWOSAMPLE-T C1 C2" returns a brief summary of the data, together with a 95% confidence interval for the difference between the means of the two variables and the calculated "degrees of freedom". For example, if we set daily temperature ranges for April, for the two locations Caribou, ME, and Albany, NY, into columns C1 and C2 of the MINITAB worksheet, the MINITAB command call returns the following information.

```
TWOSAMPLE T FOR C1 VS C2
          N        MEAN     STDEV    SE MEAN
C1        30       9.36     3.86     0.705
C2        30       6.05     3.19     0.583

95 PCT CI FOR MU C1 − MU C3: (1.480, 5.146)

TTEST MU C1 = MU C2 (VS NE): T = 3.62  P = 0.0006  DF = 56
```

Now suppose that instead of a 95% confidence interval for the difference in means, we want an interval with a different measure of confidence, such as an 85% confidence interval. Although the MINITAB command does not give this to us directly, we can construct it from the information output from

"TWOSAMPLE T FOR C1 VS C2"

and one INVCDF call for the t distribution, with the calculated degrees of freedom given in this output. An 85% confidence interval is constructed from the probability statement

$$0.85 = \Pr\left[t_{56;0.075} < \left[(\bar{X} - \bar{Y}) - (\mu - \nu)\right]/\sqrt{S_X^2/M + S_Y^2/N} \le t_{56;0.925}\right]$$

as

$$0.85 = \Pr\left[(\bar{X} - \bar{Y}) - t_{56;0.925}\sqrt{} \le (\mu - \nu) < (\bar{X} - \bar{Y}) - t_{56;0.075}\sqrt{}\right].$$

Since the t distribution is symmetric, $t_{56;0.075} = -t_{56;0.925}$. Thus we only need the value of $t_{56;0.925}$, which is output with the MINITAB command sequence "INVCDF 0.925; T WITH DF = 56." as the value 1.4596. $(\bar{X} - \bar{Y})$ is calculated from the difference of the output values of the means, here

$$(\bar{X} - \bar{Y}) = 9.36 - 6.05 = 3.31.$$

The TWOSAMPLE T value, of 3.62, for

$$t = (\bar{X} - \bar{Y})/\sqrt{S_X^2/M + S_Y^2/N}$$

permits us to calculate

$$\sqrt{S_X^2/M + S_Y^2/N} = 3.31/3.62 = 0.915.$$

Putting these components together now, we have

$$0.85 = \Pr[3.31 - 1.4596(0.915) \le (\mu - \nu) < 3.31 + 1.4596(0.915)];$$

and the 85% confidence interval for the April difference in temperature ranges for Caribou and Albany is [1.97, 4.65).

8.5 FISHER'S F DISTRIBUTION FOR A RATIO OF VARIANCES

Fisher's F distribution describes the stochastic behavior of a ratio of variance estimates under the same conditions in which we use Student's t for the comparison of means. Literally, Fisher's F distribution describes the stochastic behavior of the ratio of two, independent χ^2 variables, each divided by its degree of freedom

$$F_{\kappa_1, \kappa_2} = \left(\chi_{\kappa_1}^2/\kappa_1\right)/\left(\chi_{\kappa_2}^2/\kappa_2\right). \tag{8.18}$$

The parameters for this distribution are the degrees of freedom associated with the numerator and those associated with the denominator. By tradition, they are given in this order. To illustrate, if

$$(M - 1)S_X^2/\sigma^2 \quad \text{and} \quad (N - 1)S_Y^2/\tau^2$$

have been computed from independent

$$\mathbf{X} \to \mathcal{N}(\mu \mathbf{U}, \sigma^2 I) \quad \text{and} \quad \mathbf{Y} \to \mathcal{N}(\nu \mathbf{U}, \tau^2 I), \text{ where } \mathbf{U} = \begin{pmatrix} 1 \\ \vdots \\ 1 \end{pmatrix},$$

then they qualify; and the ratio

$$F = \left(S_X^2/\sigma^2\right)/\left(S_Y^2/\tau^2\right) \tag{8.19}$$

has Fisher's F distribution with parameters $(M - 1)$ and $(N - 1)$, respectively.

Like the χ^2, the F statistic is positive valued, and its distribution is asymmetric and long-tailed on the right. For numerator and denominator degrees of freedom κ_1 and κ_2, and an input value for P, the MINITAB command

$$\text{“INVCDF } P; \text{ F } \kappa_1 \ \kappa_2.\text{”}$$

returns the value with probability P to the left of it. We draw this as

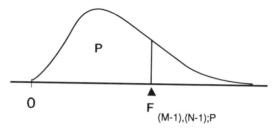

and write

$$P = \Pr\left[F \le F_{(M-1),(N-1);P} \right].$$

Construction of a Confidence Interval for a Ratio of Variances

Notice that we may rewrite (8.19) as

$$F = \left(S_X^2 / S_Y^2 \right) / \left(\sigma^2 / \tau^2 \right) \tag{8.20}$$

with the ratio of the observed variances in the numerator and the true variance ratio in the denominator. In this form we can easily see how its distribution may be used to obtain confidence intervals for the true variance ratio. By analogy with the construction of a confidence interval with a confidence value of $1 - p$, for a single variance value from a distribution, we first allocate the residual probability p to the tails of the distribution. If we want equal-sized tails, then we find $F_{(M-1),(N-1);p/2}$ and $F_{(M-1),(N-1);(1-p/2)}$, so that we may write

$$1 - p = \Pr\left[F_{(M-1),(N-1);p/2} < F \le F_{(M-1),(N-1);(1-p/2)} \right].$$

Now we substitute for F using (8.20) and rearrange the argument of the probability function, to get

$$1 - p = \Pr\left[F_{(M-1),(N-1);p/2} < \left(S_X^2 / S_Y^2 \right) / \left(\sigma^2 / \tau^2 \right) \le F_{(M-1),(N-1);(1-p/2)} \right]$$

$$= \Pr\left[\left(S_X^2 / S_Y^2 \right) / F_{(M-1),(N-1);(1-p/2)} \right.$$

$$\left. \le \left(\sigma^2 / \tau^2 \right) < \left(S_X^2 / S_Y^2 \right) / F_{(M-1),(N-1);p/2} \right].$$

Thus we have obtained a two-sided $(1 - p) \times 100\%$ confidence interval for the ratio of the true variances of the observed variables:

$$\left[\left(S_X^2 / S_Y^2 \right) / F_{(M-1),(N-1);(1-p/2)}, \left(S_X^2 / S_Y^2 \right) / F_{(M-1),(N-1);p/2} \right).$$

An Example

We will compare variances of the January, monthly means of daily average temperatures for the two periods 1980–1989 and 1930–1949, at State College, with the construction of a confidence interval for the ratio of the true variances for these periods. With two columns of data in the MINITAB worksheet, 10 values for 1980–1989 and 20 values of January monthly means for 1930–1949, the DESCRIBE command provides us with observed standard deviations and thus the observed variance values

$$(S_{1980-89})^2 = (3.90)^2 \quad \text{and} \quad (S_{1930-49})^2 = (5.51)^2.$$

For the F ratio, the numerator degrees of freedom is $M - 1 = 9$ and the denominator degrees of freedom is $N - 1 = 19$. If we choose a 90% confidence interval for the ratio of variances, and put probability 0.05 in each end of the F distribution in its construction, then the MINITAB commands that give us the F values we need are

"INVCDF 0.05; F 9 19." and "INVCDF 0.95; F 9 19."

These return

$$F_{9,19;0.05} = 0.34 \quad \text{and} \quad F_{9,19;0.95} = 2.42$$

so that we can write

$$0.90 = \Pr\left[\left(S_{80-89}^2/S_{30-49}^2\right)/F_{9,19;95} \le \left(\sigma^2/\tau^2\right) < \left(S_{80-89}^2/S_{30-49}^2\right)/F_{9,19;05}\right]$$

$$= \Pr\left[(3.90/5.51)^2/2.42 \le \left(\sigma^2/\tau^2\right) < (3.90/5.51)^2/0.34\right].$$

The 90% confidence interval for σ^2/τ^2 is $[0.21, 1.47)$. To interpret this in the context of comparing the variances of the two periods, we note that it clearly brackets the value 1.0, which is the equal variance ratio. And we can be confident in stating that the data support an assumption of equality of true variances.

   VIP: For historic reasons relating to computational and tabular efficiency, an F ratio is always formed with the observed variance having the smaller number of degrees of freedom in the numerator. Calls to the MINITAB command INVCDF for the F distribution follow this convention as well. Thus it is critical to maintain the correspondence between the variance estimates and their associated degrees of freedom, and place them in the order $\kappa_1 < \kappa_2$.

Anticipating Linear Regression

Regression analysis may be viewed as a large and complex body of statistical analysis procedures. Or it may be viewed in terms of its elementary role: as *a mechanism for allocating variability of an observed system to designated sources*. It is both. The complexity of the library of available procedures is due to differences in the manifestations and representations of physical relationships between the variables we observe and analyze, and it is due to differences in constraints placed on an analysis by the mechanisms for obtaining observations, rather than to complexity of the analysis objective. In this book we will introduce regression analysis in a relatively simple format and give you references to specializations which you may require in the future.

"Allocation of variability to designated sources" generally means that we have observed other variables in addition to the variable of primary interest; and that we wish to determine whether these other variables have significant relationships with the primary variable. Generally we make this determination with a partitioning of the total variance of the primary variable that allocates influence to "explanatory variables" and "residual variability". An *F*-ratio of components of the total variation is formed and compared with Fisher's *F* distribution. This comparison generates a measure of the credibility of hypothesized relationships among variables. Judgment of the outcome is a *test of hypotheses*. In principle this is the same as comparing the variances of data from different sources. It differs in that, in regression analysis, the estimates of components of the total variance have been computed from a single multidimensional data set which we are using to estimate "the dimension of the variance of the primary variable". This subject is of central importance in Atmospheric and Ocean Sciences. However, we will leave further exploration of the topic of regression analysis until we have completed the material on testing hypotheses, in the next chapter.

8.6 THE CORRELATION ESTIMATE AND FISHER'S Z

Thus far in the present chapter we have been concerned with inferences about means and variances, and their differences and ratios, when the observed variables have Multivariate Normal distributions. In exploring techniques for the comparison of the true means or the true variances of distinct variables X and Y, we assumed that the sets of observations of X and Y that we used in computing estimates of the true parameter differences or ratios were stochastically independent. What remain to be explored in the context of Multivariate Normal parameter estimation are the distribution of the observed correlation of a pair of variables and mechanisms for making inferences about the true correlation. Specifically, we presume here that it is the relationship between observations of X and Y which is our primary concern; and we focus on the correlation coefficient as representative of this relationship.

The Implicit Linearity of the Relationship

Evaluating the relationship between two variables with a correlation coefficient, implicitly assumes that the relationship between them can be represented by a linear function. When we discuss linear regression in Chapter 10, this implicit assumption will be examined and illustrated much more fully than the treatment we give it here. Nonetheless we wish to put into perspective an assumption which may be considered rather limiting, at first mention. The assumption that *one variable can be represented as a linear function of the other* is not an assumption which you would wish to make for most pairs of dynamically related variables in your area of science. Neither do we propose this. Rather, we apply the assumption to *the increments* between the values you observe and the values given by a dynamical model. The stochastic variables are defined as these differences, rather than as the field variables themselves. *The increments are the residuals between recorded observations of field variables and what you already know about them.* Thus known dynamical relationships of the field variables are not components of the stochastic increment data to which the linearity assumptions of correlation studies and classical regression analyses are applied.

We may explore the correlation between two increment variables in order to refine our knowledge of the relationship between the underlying field variables. However, more often we explore their correlation so that we may use what we learn about the relationship between the increments, in conjunction with dynamical models, for large-scale spatial analyses and predictions.

True Correlation and the Statistic Used to Estimate It

Recall that the true correlation between two variables X and Y, with respective means and variances,

$$\mu \quad \text{and} \quad \nu \quad \text{and} \quad \sigma^2 \quad \text{and} \quad \tau^2$$

is defined as the ratio

$$\rho = \mathscr{E}[(X-\mu)(Y-\nu)] / \sqrt{\mathscr{E}\left[(X-\mu)^2\right]\mathscr{E}\left[(Y-\nu)^2\right]} \, .$$

We may rewrite this as

$$\rho = \mathscr{E}\left[\left(\frac{X-\mu}{\sigma}\right)\left(\frac{Y-\nu}{\tau}\right)\right]$$

to see it more easily as a measure of covariation between increments which have been scaled by their individual measures of variability.

In a correlation study, the observed values of increments for the two variables come in pairs, and their "matching" must be maintained in order to study the relationship between the increments of two variables. Usually the observed values are paired in time or location, or relative to other physical circumstances. In maintaining the pairing, we are controlling for the possibility that these circumstances could influence the evidence of the relationship. We denote a data set for observed pairs from a sequence of like or coincident situations as

$$(X_1, Y_1), \ldots, (X_N, Y_N).$$

From Chapter 3, we have the formula for *the observed correlation for the data set*

$$r = \sum_{j=1}^{N} \left(X_j - \bar{X} \right)\left(Y_j - \bar{Y} \right) \Bigg/ \sqrt{\sum_{j=1}^{N} \left(X_j - \bar{X} \right)^2 \sum_{j=1}^{N} \left(Y_j - \bar{Y} \right)^2}$$

and this is a logical choice of a statistic to estimate ρ and to use as a basis for inferences about the true correlation. The numerator, divided by $(N-1)$, estimates the numerator of ρ and each sum of squares in the denominator, divided by $(N-1)$, estimates the corresponding variance in the denominator of ρ.

The distribution of the statistic r is complicated by the fact that it is a function of three, highly dependent statistics. Unlike χ^2, t, and F, we do not have a mathematically clean derivation of the distribution of r even when we are confident of the validity of the normality assumptions:

(i) $\begin{pmatrix} X \\ Y \end{pmatrix} \to \mathscr{N}(\begin{pmatrix} \mu \\ \nu \end{pmatrix}, \begin{pmatrix} \sigma^2 & \rho\sigma\tau \\ \rho\sigma\tau & \tau^2 \end{pmatrix});$

  (ii) the N pairs of observed increments are stochastically independent of one another;   (8.21)

  (iii) the five parameters μ, ν, σ^2, τ^2, and ρ, are constant; i.e., the parameters are representative of all the situations for which we have observations.

Nonetheless, with these assumptions we can rewrite our inferences in terms of a transformation of the statistic r; and use the following result, due to R. A. Fisher (see Brownlee, 1965, p. 414), as the basis for inferences about the true correlation value.

ASSERTION 8.2. *When the conditions* (8.21) *describe the vector* $\begin{pmatrix} x \\ Y \end{pmatrix}$ *then*

$$Z = \frac{1}{2}\ln\left(\frac{1+r}{1-r}\right)$$

has a distribution which is closely approximated by the Univariate Normal distribution, with mean and variance

$$\zeta = \frac{1}{2}\ln\left(\frac{1+\rho}{1-\rho}\right) \quad \text{and} \quad \eta^2 = 1/(N-3).$$

With this result we can construct approximate confidence intervals for ρ. For example,

$$0.68 = \Pr\left[-1.0 < \frac{(Z-\zeta)}{\sqrt{-1/(N-3)}} \le +1.0\right]$$

$$= \Pr\left[Z - 1.0/\sqrt{N-3} \le \zeta < Z + 1.0/\sqrt{N-3}\right]$$

$$0.80 = \Pr\left[Z - 1.28/\sqrt{N-3} \le \zeta < Z + 1.28/\sqrt{N-3}\right]$$

$$0.95 = \Pr\left[Z - 2.0/\sqrt{N-3} \le \zeta < Z + 2.0/\sqrt{N-3}\right].$$

To translate these to confidence intervals for ρ, we plot the relationship between ρ and ζ, and read off the values of ρ corresponding to the endpoints of the intervals for ζ.

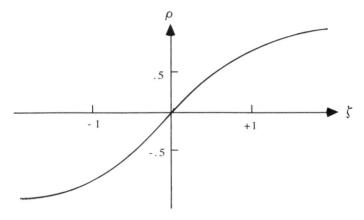

EXERCISES

1. Consider December means of daily average temperatures for three locations of our CLIMAT data set, together, in a vector of three elements: $\mathbf{X} = (X_1, X_2, X_3)^T$. For this exercise we will assume that we know the value of the ensemble mean vector

$$\boldsymbol{\mu} = (10, 0, 10)^T$$

and that the covariance matrix is

$$\Sigma = \begin{pmatrix} 2 & 0 & 1 \\ 0 & 1 & 0 \\ 1 & 0 & 1 \end{pmatrix} = BB^T, \quad \text{for } B = \begin{pmatrix} 1 & 0 & 1 \\ 0 & 1 & 0 \\ 0 & 0 & 1 \end{pmatrix}.$$

(a) Write $(\mathbf{X} - \boldsymbol{\mu})^T \Sigma^{-1}(\mathbf{X} - \boldsymbol{\mu})$ as a scalar expression, like the right-handside of (8.2).

(b) What is the distribution of this statistic? How do you know?

(c) Show the transformation to a variable $\mathbf{Y} = (Y_1, Y_2, Y_3)^T$ that has mean and covariance matrix

$$\boldsymbol{\nu} = \begin{pmatrix} 0 \\ 0 \\ -10 \end{pmatrix} \quad \text{and} \quad I = \begin{pmatrix} 1 & 0 & 0 \\ 0 & 1 & 0 \\ 0 & 0 & 1 \end{pmatrix}.$$

Show confirmation of this claim. What is the distribution of $\mathbf{Y}$?

(d) Write $(\mathbf{X} - \boldsymbol{\mu})^T \Sigma^{-1}(\mathbf{X} - \boldsymbol{\mu})$ as a scalar expression in Y_1, Y_2, Y_3.

(e) What are the distributions of (i) Y_1^2, (ii) Y_2^2, (iii) $(Y_3 + 10)^2$, and (iv) Why?

2. Refer to Exercise 1, but here assume that

$$\Sigma = \sigma^2 \begin{pmatrix} 2 & 0 & 1 \\ 0 & 1 & 0 \\ 1 & 0 & 1 \end{pmatrix},$$

where we do not know the value of σ^2.

(a) Rewrite (8.7) for this example and give the degrees of freedom for the χ^2 distribution.

(b) Find the endpoints of an 80% confidence interval for σ^2, in terms of S^2.

(c) Using the vector $(X_1, X_2, X_3)^T = (9.3, 0.7, 10.5)^T$, evaluate the endpoints. Write the fiducial probability statement corresponding to (8.7), using the endpoint values obtained here.

3. Review Exercise 1 in Chapter 7. Get the two subfiles of January mean daily average temperatures that you created there in your MINITAB worksheet: one subfile for the first 47 years' data and the other for the second 47 years' data.

(a) Construct 80% confidence intervals for the variances of these two periods.

(b) Show the confidence intervals relative to one another, on an axis of σ^2 values.

(c) Find the variance of the January means for the first 47 years and compare this with the 80% confidence interval for the second 47 years. Write a statement summarizing the evidence concerning a possible change in January mean temperature variability over the 94 years of the data record.

4. Use July data from the file 94YEARS. Make two subfiles of July mean daily average temperatures: one for the years 1896 to 1969 and one for 1970 to 1989.
   (a) Find (i) the variance of the July mean temperatures for the 74-year period beginning with 1896, and (ii) a one-sided 90% confidence interval for the true variance for the 20-year period beginning with 1970.
   (b) Write a diary note describing the construction of the confidence interval. Include an explanation of your choice of apportionment of the residual 10% of the probability. Write the expression you used to compute the endpoint statistic in terms of the statistic S^2 and the χ^2 variable. What is the value of the "degrees of freedom" for this interval?
   (c) Compare the variance for the earlier 74-year period with the 90% confidence interval for the variance of July mean temperatures for 1970 to 1989. Write two sentences giving your opinion and support for it, concerning the proposition that the variance is greater in recent years than it was earlier in the 20th century.

5. Use the TEMPS file. Make five subfiles with 37 values each: one subfile for each city, each with average Spring temperature range values for the 37 years.
   (a) Make a diary note of how you did this with MINITAB commands and include it in your homework.
   (b) For each city, find an 85% confidence interval for the true variance of average Spring temperature ranges, and plot these as

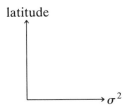

   Identify each interval with the city name written to the left of the ordinate axis.

6. Use the STCOLL file. Make a new file, JANSNW with 10 values giving total snowfall during January, for each year of 1980 to 1989.
   (a) Plot the 10 total January snowfall values versus the year.
   (b) In words, describe what you observe about the year-to-year variability of January snowfall.
   (c) Find a 90% confidence interval for the variance of total January snowfall. Write the mathematical expression for this interval; give

the values of each of its elements; and write the final interval in terms of a fiducial probability statement.

(d) Translate the fiducial probability statement into a verbal statement.

7. Review Exercises 2.5 and 7.7.
   (a) Use Boise temperature range values for the first 60 days of 1990. (i) Find an 80% confidence interval for the true mean of the observed temperature ranges. (ii) Write the formula you use to compute the endpoints of the interval and identify each element with its value. (iii) Write the probability statement corresponding to the 80% confidence interval.
   (b) Write the verbal statement equivalent to the latter probability statement.
   (c) What assumptions have you made in constructing the interval?
   (d) If the assumptions are demonstrably untrue, what must you conclude about the interval you have computed?

8. Use the JANSNW file created in Exercise 8.6.
   (a) Find a 90% confidence interval for μ, for January snowfall at State College, using the data for the decade of the 1980s.
   (b) Write a "diary note" stating how you have computed the interval endpoints.
   (c) What assumptions have you make in constructing the interval? Using what you know about annual variation in snowfall, do you think your assumptions are valid? Justify your answer.

9. Use the STN#27 file.
   (a) Make a small file with two columns: one column containing the 12 differences between salinities at 10- and 100-m depths, and a second column containing the 12 differences between temperatures at 10 and 100 m. Print and label the file, and include it in your homework.
   (b) Give the mathematical expression for an 85% confidence interval for a true mean differential, which may be applied to either salinity or temperature. Identify the degrees of freedom.
   (c) Find the 85% confidence intervals for: (i) the true mean salinity differential; (ii) the true mean temperature differential.
   (d) Why is the question of equality of the variances of the salinity and temperature variables at 10 and 100 m not an issue in calculating the confidence intervals?

10. Use the CLIMAT file.
    (a) For the data for the Winter of 1989–1990, find a 90% confidence interval for the daily average pressure differences between Nome

and Fairbanks. Write a diary note for the steps in the data selection and computation.

(b) With the data for the Summer of 1990, find a 90% confidence interval for the daily average pressure differences between Nome and Fairbanks.

(c) Compare the intervals you obtained in (a) and (b). State whether you think there is evidence that Winter and Summer pressure increments differ. Support your conclusion.

11. Use October 1989 and May 1990 daily peak wind speed data for Charleston, from the CLIMAT file.

(a) Assume equal variances for the daily peak wind speeds of these two months, and construct an 80% confidence interval for the difference between their two means.

(b) Construct an 80% confidence interval for the difference in true means without assuming that peak wind speeds have identical variances in these two months.

(c) Comment on how these intervals differ. If you wish to be on the safe/conservative side, which would you use? Explain your choice. What is its disadvantage?

12. Review Exercise 8.3. Retrieve the data you used for that exercise.

(a) Construct a 95% confidence interval for the difference between January mean daily average temperatures for the two 47-year periods.

(b) Describe the method you used; explain your choice of methods; and identify the values of the components of the calculation.

(c) Is the value 0 in this interval? What does this suggest to you? Explain.

13. Use the STN#27 file, and temperature and salinity data for the same two levels as for Exercise 8.9.

(a) Find an 85% confidence interval for the ratio of variances of salinities at these two levels.

(b) Repeat (a) for temperatures.

(c) Verbally state your results and a comparison of them.

(d) If you have knowledge of the physical basis for the differences, explain this.

(e) What assumptions have you made in constructing the confidence intervals. If you have reason to doubt the validity of either of them, explain.

14. Use the same data as in Exercise 8.4.

(a) Calculate a 90% confidence interval for the ratio of variances for the two periods 1970 to 1989 and 1896 to 1969. Give the formula for the endpoints of the interval; and identify the values of their elements.

(b) Is the value 1.0 in the 90% confidence interval? What does this say to you?

(c) Does the evidence produced here concur with the opinion you stated in Part (c) of 8.4? Comment on this.

15. Use the TEMPS file.

(a) (i) Make a new file using just July data. Put the year in column 1, day of month in column 2, and daily maximum July temperatures in columns 7 through 11 (one column for each city), naming these last five columns with the city names. (ii) Calculate the four observed correlations of daily maximum July temperatures of State College, with those of the other four cities. (iii) Construct 80% confidence intervals for the four true correlations. Write a diary note explaining how you obtained these.

(b) (i) Create five new columns of 37 elements each, containing maximum July temperatures for the five cities: one maximum for each year, for each city. (ii) Calculate the four observed correlations: maximum July temperatures of State College, with those of the other four cities. (iii) Construct 80% confidence intervals for the four true correlations. Write a diary note explaining how (why) these differ from those you obtained in Part (a)(iii), above.

(c) Make two copies of a map that shows the locations of the five cities: Crookston, MN, Little Rock, AK, Saginaw, MI, State College, PA, and New York, NY. (i) On one of these print the pair of correlation values obtained in Parts (a)(ii) and (b)(ii), close to the location of each city, putting the correlation values for the complete July records first and the values for the yearly July maximums second. (ii) On the other map, print the pair of 80% confidence intervals for true correlations, obtained in Parts (a)(iii) and (b)(iii), close to the location of each city. put the confidence intervals for the complete July records above and the intervals for the yearly July maximums below. (iii) Look critically at the comparisons you have created. Comment on the differences in the pairs of values. Include in your comments any scientific insights which explain the differences.

16. Use the CLIMAT file and the data selected for Exercise 8.10.

(a) Find a 90% confidence interval for the correlation between daily average pressures for Nome and Fairbanks.

(b) Write a diary note describing how the confidence interval was determined.

(c) Is your conclusion that the pressures at these locations are highly correlated? Present a short argument in support of your answer.

9

TESTING HYPOTHESES
"dealing with the generic critic
while establishing powerful
support for new ideas"

9.1 INTRODUCTION

"Testing hypotheses" means using observations to determine whether we have sufficient evidence to argue that a research hypothesis is true. The research hypotheses we will discuss here are statements about ensemble characteristics of variables of an observed system, and relationships between the variables. The antithesis of a particular research hypothesis, which encompasses all the possibilities about the characteristic or relationship not specified by this hypothesis, plays a major role in testing hypotheses, because the job is to determine whether we can clearly distinguish between the research hypothesis and its antithesis. Objectively this is something of a gambling game. However, it's not a crap shoot. It's more like playing a game of blackjack against "the house", when it's possible to discover the strategy used by the house in deciding whether to "hold" or deal itself another card. What this means is that, while we cannot eliminate uncertainty and the possibility of an erroneous judgement, we *can* formulate our strategy with knowledge of the probabilistic rules associated with both the research hypothesis and its antithesis. Then, when we have specific observations and a value for whatever statistic we are using to summarize them, we use our knowledge of the probabilistic rules to make an informed decision about our beliefs in these competing hypotheses, with known risk of error.

In this chapter we focus on the details of strategies used in confronting competing hypotheses with the truth of observations and in deciding whether the antithesis of a research hypothesis can be rejected on the outcome of the confrontation. We will look closely at the role of the probabilistic rules in testing hypotheses and at the steps taken in assessing the relative credibilities of hypotheses. These are the basic components of scientific inference. Familiarity with them will be important, not only to interpreting results of your own research but also to interpreting reports given by others, in the scientific literature. Together they provide a critical filter for all observational evidence that bears on answering scientific questions.

Before looking at an example, let's consider a generic case. Suppose we are collaborating on a research project designed to establish a relationship which has not previously been known or, at least is not generally accepted as fact. The newer the concept the more exciting the project and the more promise it holds for our future reputations as scientists. We are engaged in the research because we believe we can generate evidence that will lead to acceptance of the relationship as fact. It will be exciting if we do generate strongly supporting evidence. We may even find ourselves in the scientific limelight. Clearly we will wish to be confident of the conclusions we reach from our research, when we present and defend them. As with all fresh reports of research, the results we place before the scientific community will receive critical review; and it would be unwise to make public statements which claim credit for a bit of new knowledge or scientific insights, if these were only weakly supported. Thus we rely on universally established mechanisms for testing hypotheses that expand the limits of knowledge: mechanisms which have developed from the requirements for making confident scientific inferences.

It will greatly assist you in identifying with the objectives of this chapter to project your imagination into the scientific arena as a participant in the process of new discovery. If you have an example from your past experience or something that you are working on now, carry it along in your mind as you read and think about the material of this chapter. What you read here will provide a sound foundation for the process of using data to evaluate relative credibilities of competing hypotheses. It will also provide a framework for presenting evidence in support of a specific research hypothesis. While a significant amount of attention to detail goes into creating the structure for valid hypothesis testing, the end result is certainly worth the investment when it is measured in units of inferential validity and scientific reputation.

9.2 STATEMENT AND TESTING OF COMPETING HYPOTHESES

In this and the following sections of the chapter we will meet and illustrate through examples the nomenclature and conduct of tests of hypotheses. The emphasis will be on creating a solid case for a research hypothesis in which we believe strongly. (Future career opportunities or, at the very least, future research funding may be at stake!) The requirements are: (i) that we bring sufficient evidence to bear on the issue posed by competing hypotheses, in order to make a confident decision regarding the validity of one or the other of the hypotheses, and (ii) that we present an indisputable defense of our scientific conclusion. In this connection we will introduce and discuss the concepts of *the credibility of a hypothesis relative to the observations* and *the power of a statistical test;* and we explore the relationship of the latter to the number of observations that are used in a decision process. We will make the important inferential connection between confidence intervals for parameters and tests of hypotheses about their values. Finally, we will look at the problem of using statistics for which our knowledge of the observed system does not give us definitive distributions. The latter leads into the topic of the final chapter of the text.

Designation of the Competition

The rationale for the structure of hypothesis testing has as its source the assumption that objective critics will doubt the validity of any newly proposed relationship. Thus the point of hypothesis testing is to challenge research results with the possibility that what we are seeing is purely a consequence of chance variations among values of the variables under observation, operating in the absence of the relationship proposed by the research hypothesis. We cannot absolutely dismiss this possibility. However, we can develop a mechanism for assessing the probability of obtaining a research outcome which is as strongly in support of the proposed relationship as ours is, if in fact this relationship is *not* operative. (Thus we assign a likelihood to the leg that a potential critic might stand on.) We do this by first articulating a counter hypothesis, which might be called "the skeptic's hypotheses". It is the antithesis of the research hypotheses. Then what we seek to do is to establish an infalible argument for rejecting it. In the context of bringing to light new scientific knowledge, accepting the skeptic's hypotheses as true would be a totally uninteresting outcome for a research project—in agreement with our desire to reject it. Traditionally this *antithesis* is labeled *the null hypothesis* and denoted by H_0. The tradition will serve us, because it is descriptive in any setting. However, rather than relegating the focus of our research program to the traditionally anonymous "alternative hypothesis" designation, we will call it as it is, namely *the research hypothesis* and denote it by H_R.

As an example for discussion of testing hypotheses, consider the question of whether there was a climate warming in North America, during the latter part of the twentieth century. Those who believe there was, put forward a research hypothesis stating their conjecture. Since the skeptic's hypothesis, or null hypothesis, is the antithesis of the research hypothesis, it includes "no change" as well as the possibility of a cooling. Of course both hypotheses must be phrased in terms of available evidence, namely the existing data and statistics relevant to answering the question posed by the competing hypotheses, whose values can be calculated with the data. For our example we will use the July monthly means of daily average temperatures which were derived from the 94YEAR data file for State College. In the context of this data set, the research hypothesis can be phrased as

H_R: The true mean of July temperatures was greater for the period 1950 to 1989 than the true mean for the period 1896 to 1949

and the null hypothesis as

H_0: The true mean of July temperatures for the period 1950 to 1989 was either the same or less than the true mean for the period 1896 to 1949.

Thus we have defined the competition in terms of the information that is available to carry it through.

The Mechanism for Selecting between Competing Hypotheses

After designating the competing hypotheses, the next step is the formulation of a decision procedure, or rule, which will reject the null hypothesis H_0 if the evidence conveyed by the observations strongly favors the research hypothesis. The decision procedure is formulated around the condensation of the information in a (generic) set of observations, that is provided by a summary statistic: a statistic chosen to summarize the information in the data which is most relevant to establishing the validity of H_R. We call this *the test statistic*. Ideally it is *the algorithm giving the optimal condensation of the information in the observations of the system under study, relative to the objective of making a confident decision regarding the validities of the null versus research hypotheses*. It may be as simple as a difference of observed means, a ratio of observed variances, an observed correlation coefficient, or the slope of a regression line, or it may be a combination of parameter estimates, such as the normalization of the difference between the means of two sets of data provided by Student's t.[1] Whatever the relevant test statistic is, the test of H_0 versus H_R will use it as the "bottom line discrininator". The structure of the test of hypotheses will be straightforward if we have chosen as our test statistic a summarization of the data which clearly mirrors the distinction between the hypotheses.

The two primary keys to testing hypotheses are:

(i) Selection of a test statistic which is sensitive to the difference between the hypotheses H_0 and H_R;
(ii) Clarity about the appearance of evidence in support of each hypothesis, as likely values of the test statistic.

In the barest terms, what we are going to do with these keys is to partition the collection of all possible sets of observations into two: one corresponding to all those that could lead to rejection of the null hypothesis and the other corresponding to all those for which we could not reject it (either because it is manifestly true or because we do not have a sufficient number of observations to establish unequivocally that it is not true). *Every construction of a test of hypotheses is a simple bifurcation*, which we ask you to identify more with cutting a diamond than with chopping kindling.

Setting up a null hypothesis with the object of rejecting it may seem like a counter productive expenditure of intellect and other valuable resources. However, our objective is a resounding dismissal of the possibility that apparent support for the research hypothesis is purely a figment produced by chance variations. To achieve this objective, we first acknowledge that different observers or observing equipment, or small-scale variations in the ocean or atmosphere which are of no relevance to our research hypothesis, would produce somewhat different observations and correspondingly different values of the test statistic. Next we identify (or generate) the distribution of the statistic which would pertain if the null hypothesis were operative. This distribution reflects all the aforementioned irrelevant sources of variability acting in concert with the statement of H_0. Using this distribution we evaluate the collective probability of all the values of the test statistic possible under H_0, which would favor the research hypothesis at least as much as the value we actually obtained. This collective probability is a good measure of just how extreme the observed value of the test statistic is relative to the null hypothesis; or, by inverse argument, it is a measure of *the credibility of H_0 relative to the*

---

[1] In atmosphere and ocean sciences this is frequently compounded by having a time- or space-indexed array of statistics to consider at once, in addressing a research hypothesis. This situation is another level of complexity, and will be taken up in the final chapter.

observations.[2] If this probability is small we can confidently reject the null hypothesis, because the probability is predicated on H_0. In this event we will have established that *it is very unlikely that the situation described by the null hypothesis could have produced an outcome as extreme as ours, in the direction favored by* H_R; *and we will choose to accept* H_R. If the credibility of H_0 relative to the observations is not sufficiently small, then we cannot reject H_0 and accept H_R, at least not until we accumulate more support from additional observations.

In an ideal situation we could evaluate the collective probability predicted on the relationship described by the research hypothesis, corresponding to the credibility of H_0. Then we would have *relative likelihoods of the set of possible outcomes which favor* H_R *at least as much as the outcome we obtained favors* H_R: one based on H_0 and the other on H_R, on which to base the decision procedure for choosing between the hypotheses. However, this is almost never possible. Evaluation of the relative credibility of H_R requires use of the distribution of the test statistic that pertains when the research hypothesis is valid; and generally this distribution is not uniquely specified. To see this, think about the research hypothesis: "Total precipitation during winter is related to the average temperature of the preceding summer, where we take 'related to' to mean that there is a non-zero correlation, ρ, between winter precipitation and summer average temperature." Assume that we have an archive with many years of data for these variables, say P_{Winter} and T_{Summer}. A test statistic which may be calculated from this resource is the observed correlation which can be calculated from values in the archive. Now we write a specific null hypothesis as

The true correlation between P_{Winter} and T_{summer} is 0.

We know the distribution of the test statistic if H_0 is true. However, when we try to specify the distribution of the observed correlation coefficient when H_R is true, we're stuck. The research hypothesis does not specify the value of the true correlation, except to say that it is not zero. It does not even tell us whether ρ is positive or negative! Thus H_R corresponds to a double continuum of possible distributions for the test statistic, with an associated relative credibility for each distribution.[3]

---

[2] Traditional statistical terminology for this probability is "the significance of the test statistic". However, this designation is not as descriptive of its role in the scientific decision process as calling it a credibility.

[3] Even when we select a specific non-zero value for ρ, evaluation of the credibility of H_R relative to the observations presents a major challenge, because the analytic form of the distribution of a correlation coefficient calculated from pairs of stochastically dependent observations is not known. The distribution could be created by Monte Carlo simulation, as discussed in the final chapter. However, this requires a major effort which is usually not made in hypothesis testing, since it would assign relative credibility of only one of the continuum of possibilities which comprise H_R.

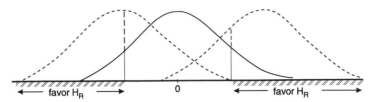

FIGURE 9.1   The curve drawn with the solid line represents the density function for the Z-transformation of observed correlation when $\rho = 0$. Dashed and dotted curves represent density functions of Z for specific values of $\rho \neq 0$.

The credibility of H_0 relative to the observations (cH_0ro) does not bear any necessary relationship to the relative credibility of any of the research alternatives, except that it will always be smaller. Thus we generally use the magnitude of cH_0ro alone, in reaching a final decision regarding the competing hypotheses. Classic "small values" of CH_0ro for which folks dismiss the null hypothesis with impunity are 0.05 and 0.01. However, these choices are arbitrary. Once you feel comfortably familiar with the structure of the decision-making process and the playoffs involved in testing hypotheses, you will want to make your own selections. And they may well be different in every testing situation. When in doubt, the cH_0ro can be reported without reference to a "cut off value" and the final decision left as an individual choice. In fact, more information is conveyed by a research report that gives the exact value of the credibility of H_0 relative to the observations.

We can generally characterize the *locations* of the H_R distributions relative to the H_0 distribution, in terms of the test statistic, even though we may not know their exact forms. We use our knowledge of relative locations to designate which of the possible values of the test statistic should lead us to reject H_0 and accept H_R; and we call this set of values *the H_0 rejection region*. Logic dictates that the values in the H_0 rejection region be relatively much more likely when the research hypothesis is true. In the illustration provided by Fig. 9.1, this region is the union of two intervals: those values far to the left of 0 and those far to the right of 0. Choice of the limits that specifically define a rejection region, such that we can confidently declare that the research hypothesis is true when the value we obtain for the test statistic is beyond one of the limits, is the crux of hypothesis testing. Factors affecting these choices are considered below.

An Example

We can use our question about the correlation between average winter precipitation and the preceding summer's average temperature to explore the choice of intervals of values of a test statistic for which we would feel confident in rejecting a null hypothesis, so that we can accept a research hypothesis. Our investigation is "exploratory"; that is, we don't have a prior idea about whether the hypothetical correlation is a positive or negative one. Thus our research hypothesis is

P_{winter} and T_{summer} have non-zero correlation.

In this case, the antithetical or null hypothesis is

P_{winter} and T_{summer} have correlation equal to 0.

We are saying that we wish to detect any linear relationship between these two variables, whether it is that higher (lower) than normal average winter precipitation is generally preceded by a summer with higher (lower) than normal average temperature, implying positive correlation, or by a summer with lower (higher) than normal average temperature, implying negative correlation. Accordingly the test will require a two-ended rejection region for H_0: one interval corresponding to correlation values significantly greater than zero and the other to correlation values significantly less than zero.

Since we will base this test on the correlation, we should use Fisher's Z-transformation of the observed correlation coefficient

$$Z = \left[\log(1 + r)/(1 - r)\right]/2$$

We know its distribution when the null hypothesis is true; so we can evaluate the probabilities of intervals of possible observed values. Under H_0 this distribution is approximately Normal with mean 0 and variance $1/(N - 3)$. On the other hand, the research hypothesis simply says that the mean is *not* 0: it might be positive *or* negative. Furthermore, if H_R is true, then we do not know the exact form of the distribution. We cannot evaluate probabilities of intervals, but only sketch distributions relative to the known form and location of the null hypothesis distribution of Z. Visualize the latter as the solid curve in Fig. 9.1 and the dashed and dotted curves as possible distributions of the test statistic if the research hypothesis is true, i.e., if $\rho \neq 0$. Here a decision rule based on Z, that will reject H_0 if we have evidence strongly favoring either of the possible alternatives shown, must reject H_0 for either large positive or large negative values of the test statistic. The H_0 rejection region is two-ended, as indicated in the figure by the cross-hatched regions of the abscissa.

9.3 THE PLAYOFF BETWEEN RISKS OF WRONG DECISIONS

When we choose limits for the set of possible values of the test statistic for which we will reject H_0, we automatically set *the size of the rejection region*. In the example given, this is the sum of the probabilities of the left and right intervals, where these probabilities are evaluated under the assumption that H_0 is true. In general it is *the total probability that the test statistic would have a value in the* H_0 *rejection if the null hypothesis were indeed true.* This is the probability of making the first kind of wrong decision: *rejecting* H_0 *when it is true.* Clearly we wish to choose interval endpoints so that this probability will be small. At the same time, we want the intervals to have relatively large probabilities of containing an observed value of the test statistic when the research hypothesis characterizes the true state. Look at Fig. 9.1 and think about moving the interval endpoints out: further away from zero, in order to decrease the probability of rejecting the null hypothesis when it is in fact true. If we do that, we also decrease the probabilities of getting an observed value of the test statistic in the H_0 rejection region when important non-zero correlation values actually pertain. Thus we are confronted with the necessity of articulating a decision rule which accommodates this "playoff of risks" in a reasonable way.

In classical statistical terminology the two kinds of wrong decisions, namely *rejecting* H_0 *when in fact it is true* and *not rejecting* H_0 *when the research hypothesis is true*, are called *Type I and Type II errors*, respectively. If you recall this tradition you may also recall that the probabilities of these errors of judgment, which we wish to minimize in our construction of the decision rule, are given the Greek letter designations α and β. We have just observed that there may be a tension between them: making one smaller makes the other larger. This is generally the case; and the final compromise is just that: a compromise. It is important that we reach it with the fullest possible information about its consequences vis-à-vis our research objectives.

An exploration of the playoff of the risks of making wrong decisions with our hypothesis-testing algorithm will be simpler if we agree to use a *generalized research parameter* θ to index all possibilities encompassed by the hypotheses. This parameter may be vector valued or it may be one specific (scalar) parameter of an otherwise known distribution. In any case, we will regard it as *an index of all the possible distributions of the test statistic*, under the combined null and research hypotheses. Using this parameter, we will look at testing hypothesis from the objective viewpoint of wanting our work to substantiate the research hypothesis for a particular value of θ which we will denote as θ_1. Let us assume that θ_1 is sufficiently different from the value ascribed by H_0, say θ_0, that it is important that the test of hypotheses reject H_0 when θ_1 is the true research parameter. Its importance focuses our attention on *the probability that the observed value of the test statistic is in the* H_0 *rejection region when* θ_1 *characterizes the distribution of the test statistic.* This probability is called the *power of the test for* $\theta = \theta_1$ and is denoted by $\Pi(\theta_1)$.[4] We write the power in terms of the research parameter to draw attention to the fact

---

[4]Ordinarily we will not be able to write the power as an analytic function of the research parameter, although computer-aided evaluation can always find particular values for us.

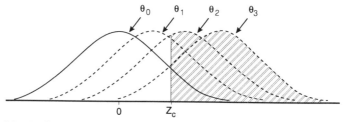

FIGURE 9.2 A diagrammatic representation of possible distributions of a test statistic, indexed by θ_0, for the null hypothesis, and $\theta_1, \theta_2, \theta_3$, as research hypothesis alternatives.

that its value will be different for each different value of θ. See Fig. 9.2 for an illustration of this, where the power for each of the possible research hypothesis distributions is the area under the corresponding curve, to the right of Z_c. We note that for every θ, the power of the test is exactly $1 - \beta(\theta)$, since we do one of two things: reject H_0 when θ is the true research parameter, or fail to reject H_0.[5]

We can diagram the hypothesis testing as follows:

State of the World

		H_0 is true	H_R is true
	Reject H_0	Type I Error α	Decision matches state $\Pi(\theta) = 1 - \beta(\theta)$
Decision:	Do not reject H_0	Decision matches state $1 - \alpha$	Type II Error $\beta(\theta)$

As represented in the diagram, there are four possible state/decision combinations. Whether the true state of the world is described by the null hypothesis or by the research hypothesis, we can make one of two decisions: either to reject H_0 or not to reject H_0. One of these correctly corresponds to the true state; the other would be an "error of judgment". If H_0 describes the true state of the world and we reject H_0, then we have made a Type I Error; if we do not reject H_0 then our decision corresponds to the true state. The situation with respect to the decisions represented here is reversed if H_R describes the true state: rejecting H_0 (and therefore

[5]The possibility of failing to reject H_0 when H_R is true highlights the importance of using the dichotomy "reject" and "not reject", with reference to H_0, rather than "reject" and "accept". If we "accept H_0" then we paint ourselves into a corner, so to speak; because that is an acceptance of the antithesis of our research hypothesis. "Not rejecting H_0" leaves open the interpretation that we have not yet assembled sufficient evidence for confident acceptance of the hypothesis to which we have dedicated our considerable efforts.

accepting H_R) corresponds to the true state, while not rejecting H_0 means that we have committed a Type II Error. For a given decision algorithm based on a test for which we know or can simulate the distribution, the probabilities of the decision options can be assessed for each research parameter. Their values are shown in the body of the above Decision: State-of-the-World table. We note that for each description of the true state of the world, the decisions are dichotomies, with probabilities that add to 1.

9.4 THE DEPENDENCE OF POWER ON THE NUMBER OF OBSERVATIONS

It may be that for some parameter values the power of the test is not large enough to satisfy us. Say for instance that, for the correlation example, it was scientifically important to accept the research hypothesis for a true correlation value of 0.6 or greater, in contrast to a null hypothesis value of 0.0, but that we have ascertained that $\Pi(0.6)$ is only 0.275 for the research program as it has been planned. This is discouraging. However, we are not without recourse if we have discovered this during the planning stages of the field work and have some flexibility in the research program. To show how this works, we look again at the distribution of the observed correlation between two variables.

We note that the variance of the test statistic is $1/(N - 3)$ when H_0 pertains; and that it will be proportional to this when any of the research hypotheses characterize the true state. Thus, if we can increase the number of observations in our research plan, we decrease the variance of the test statistic. Compare Figs. 9.1 and 9.3 to see what this does to our ability to discriminate between hypotheses: The smaller the variance the more the distribution of the test statistic is concentrated around its mean

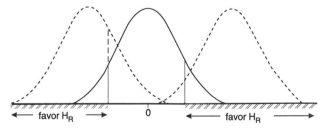

FIGURE 9.3 The curve drawn with the solid line represents the density function for the Z-transformation of observed correlation when $\rho = 0$. Dashed and dotted curves represent density functions of Z for specific values of $\rho \neq 0$. Here the number of observations is almost twice that for the representation of relative distributions in Fig. 9.1.

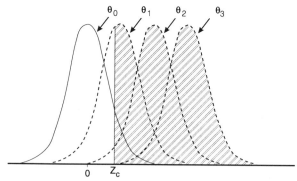

FIGURE 9.4   A diagrammatic representation of possible distributions of a test statistic, indexed by θ_0 for the null hypothesis and $\theta_1, \theta_2, \theta_3$, for research hypothesis alternatives. Here the number of observations is almost four times that for the representation of the distributions shown in Fig. 9.1.

value, whatever that mean value is, and the greater is the separation of the distributions we wish to be able to distinguish. Now compare Figs. 9.2 and 9.4, to observe the change in the powers of the same-sized test, for the same set of research parameters, with the increase in the numbers of observations that go into evaluation of the test statistic.

We should not give up on our research hypothesis unless the assembled evidence makes it clear that the null hypothesis *is* valid. If it is possible to enlarge the data set, the test statistic can be recalculated using the larger amount of information. This will increase the power of the test, because both H_0 and H_R distributions will reflect the larger number of observations, with reduced variances, and therefore greater separation.

First Illustration of Effects of Increasing Numbers of Observations

For the first illustration we choose a test of hypotheses about two mean values, such as true mean temperatures for different locations or different time periods, and investigate the power of the test for a difference in means of half a standard deviation. We'll assume that we have data sets available for the comparison, with possibly unequal numbers of observations in the two sets; and we will denote these numbers of observations by M and N. For example, we may wish to compare mean temperatures for a given month, in a recent decade, to the mean temperatures for the same month, in prior decades for which we can assemble data.

The hypotheses for which we will construct the test are

H_R: The true mean temperature for this month, during the recent decade, is higher than the true mean for preceeding decades.

and

H_0: The recent mean temperature is no greater than the prior true mean.

The construction of the test and the evaluation of its power for the specific alternative of concern require that we (i) select a test statistic, (ii) identify a region of its possible values for which we will reject H_0, and (iii) evaluate the probability of the test statistic being in this region if there has been an increase in the true mean of the observed temperatures, of half a standard deviation.

To simplify the illustration we will assume that the variances for the two periods are the same and that we know this value, so that we can use the Normal variable

$$U = \frac{\bar{X} - \bar{Y}}{\sigma\sqrt{1/M + 1/N}}$$

as the test statistic. If the true means are not different, the expected value of the numerator is zero and the distribution of U is $\mathcal{N}(0,1)$. To complete construction of the test of hypotheses, we must choose a value for α: the probability of rejecting H_0 when in fact it is true, or the corresponding *critical value* for the test statistic U_c. This is *the value for which our decision rule says*, "Reject H_0 if the observed value of U exceeds U_c." Once we have the latter we evaluate the power of the test. Specifically, the probability we wish to assess is

$$\Pi = Pr\left[\frac{\bar{X} - \bar{Y}}{\sigma\sqrt{1/M + 1/N}} > U_c | \mu - \nu = 0.5\sigma\right].$$

To the right of the vertical bar we have written the condition under which we want the value of the probability. Hence, this is the probability of "detecting an increase of 0.5σ in mean temperature", in the sense that we will reject H_0 and accept H_R if the observed value for the test statistic exceeds U_c.

We can rewrite the inequality, by applying the condition to standardize the left-hand side; and it then becomes

$$\Pi = Pr\left[\frac{(\bar{X} - \bar{Y}) - (\mu - \nu)}{\sigma\sqrt{1/M + 1/N}} > U_c - \frac{0.5}{\sqrt{1/M + 1/N}}\right].$$

Since we can scale both $\bar{X}$ and $\bar{Y}$ variables, by dividing by the standard deviation which we have assumed we know, we will simply substitute 1 for σ. Now, in terms of the standardized Normal variable Z, the formula for

TABLE 9.1   Probabilities for Detecting a True Change in Monthly Mean Temperatures as Small as 0.5σ for Some Combinations of Values of α and of M and N

	α		
M, N	0.05	0.10	0.15
10, 10	0.300	0.436	0.468
20	0.362	0.504	0.599
30	0.392	0.536	0.629
40	0.409	0.552	0.644
50	0.420	0.564	0.655
20, 50	0.595	0.726	0.800
80	0.638	0.767	0.832

the power of the test is

$$\Pi = Pr\left[Z > U_c - \frac{0.5}{\sqrt{1/M + 1/N}} \right]$$

which we can evaluate for any choice of σ and possible combinations of numbers of observations for the two periods. Table 9.1 gives a selection of combinations in which we can see the progression toward larger values of the power, for $\mu - \nu = 0.5\sigma$, as we increase the values of M and N.

The top portion of the table corresponds to values of no more than the number of years in the decade for M, with values up to 50 for N, with maximum power of 0.655. If that is not sufficient for our scientific objectives, then we need to redefine the comparison, so that we can include more years in the first data set as well as the second. This will work provided the data can be made available and provided that we do not suspect that any significant change occurred during the earlier decades. As the lower portion of the table shows, this will bring us major increases in the probability of detecting what we may very well consider an important change.

If we reflect on this illustration for a moment, it is clear that a failure to reject a null hypothesis of "no change" or "no increase'" with a relatively small number of observations should not be interpreted as "accepting this hypothesis". What we have demonstrated here is that the research design and the structure of the test may not give us sufficient power to detect an important research alternative: It may just be that we have not assembled enough evidence to accept our research hypothesis. This new tool gives us the capability of determining whether we can achieve our objective with the original plan, or whether it must be redesigned before we invest further time and resources. If the initial plan for a project is too limited for it to be likely to fulfill its objectives, it's not good science!

Second Illustration of the Dependence of Power on the Numbers of Observations

Here we choose a test for differences in variances, such as might be used in a study of the observed variances in average wind speeds for different potential power generation sites; and we evaluate probabilities for detecting a true variance ratio that differs from 1. We will assume that we have data for the two sites, such as the observations in the CLIMAT file, for which we can compute values of observed variances and use the variance ratio statistic

$$F_{(M-1),(N-1)} = \left(\frac{S_X^2}{\sigma^2}\right) \Big/ \left(\frac{S_Y^2}{\eta^2}\right)$$

to test competing hypotheses. As in most scientific projects, we wouldn't be this far along in our thinking if we didn't already suspect a difference. In fact we have probably already had at least a cursory look at the data and can identify the site for which we think the average wind speed has the greater variance. We'll identify this with S_X^2 and σ^2, and the site with the smaller observed variance with S_Y^2 and η^2, and let M and N denote the numbers of observations from the two sites that have gone into the calculations of S_X^2 and S_Y^2.

The hypotheses we wish to test are

$$H_0: \sigma^2 = \eta^2 \qquad \text{versus} \qquad H_R: \sigma^2 > \eta^2.$$

If the null hypothesis is true, the test statistic is

$$U = S_X^2/S_Y^2$$

which we assume has Fisher's F distribution with numerator and denominator degrees of freedom: $(M-1)$ and $(N-1)$, respectively. We will want to reject H_0 if the observed value of U is large relative to this distribution, indicative of a true variance ratio exceeding 1. Thus we select values for α and find the corresponding critical values for the observed variance ratio, U_c, such that we would wish to reject H_0 and conclude that H_R is true if we find $U > U_c$. Having established a decision rule in this way, we can assess the probability of detecting a true variance ratio of specified value, where again "detection" is interpreted as a rejection of the null hypothesis when truly $\sigma^2 \neq \eta^2$, as specified.

To determine the probability of detecting $(\sigma/\eta) = 1.5$ with the decision rule, we must evaluate

$$Pr\left[(S_X^2/S_Y^2) > U_c \big| (\sigma^2/\eta^2) = (1.5)^2 = 2.25\right].$$

TABLE 9.2   Critical Values for a Variance Ratio and Corresponding Powers of Tests for Equality of Variances U and Π, When the True Variance Ratio Is 2.25, for Combinations of Numbers of Observations and Significance Levels

M, N	α					
	0.05		0.10		0.15	
5, 10	3.33	0.28	1.12	0.41	0.93	0.50
10, 10	2.98	0.33	1.03	0.48	0.86	0.58
20	2.35	0.44	0.86	0.58	0.76	0.67
30	2.16	0.49	0.81	0.62	0.72	0.70
40	2.08	0.52	0.78	0.64	0.70	0.72
50	2.03	0.54	0.77	0.66	0.69	0.73
20, 50	1.78	0.75	0.70	0.81	0.64	0.86
80	1.70	0.75	0.67	0.84	0.62	0.89

Again we use the condition given to the right of the vertical bar to rewrite the inequality in terms of the variable that has Fisher's F distribution when the research hypothesis pertains

$$F_{(M-1),(N-1)} = \left(\frac{S_X^2}{\sigma^2}\right) \Big/ \left(\frac{S_Y^2}{\eta^2}\right) = \left(S_X^2/S_Y^2\right) \Big/ 2.25.$$

Now, with the parameter θ denoting the true variance ratio, the power of test of hypotheses is

$$\Pi(2.25) = Pr\left[F_{(M-1),(N-1)} > U_c/2.25\right]$$

and we can determine its values for a combination of values for α and for M and N, with repeated calls to MINITAB's CDF and INVCDF for the F distribution. Table 9.2 shows the power dependence on M and N.

9.5 THE CONNECTION BETWEEN CONFIDENCE INTERVALS AND TESTS OF HYPOTHESES

You will recall that a confidence interval for a parameter that's of interest to a research program is an interval of possible values for the parameter with an associated likelihood that this interval contains the true value. The "associated likelihood" is a probability that you, the researcher, pick. Generally its choice involves a compromise between how closely you want the parameter to be bracketed and how certain you wish to be that you have indeed "captured" the true parameter value: The smaller the interval, the less the likelihood that it includes the true value. The values of the endpoints of the interval will be calculated from observations. Thus they are random variables and their values will be different for data obtained at different times or locations, or with the use of different observing equipment. A very simple illustration, to freshen recollection, is provided by the construction of a confidence interval for the mean of a variable assumed to have a Univariate Normal distribution, when there are N stochastically independent observations at hand, say $X_1, \ldots, X_N$. Since

$$1 - \alpha = Pr\left[-t_{\alpha/2} < \frac{\overline{X} - \mu}{\sqrt{S^2/N}} < +t_{\alpha/2} \right]$$

then, by simple algebraic rearrangement of the argument of the probability, we get

$$\left[\overline{X} - t_{\alpha/2}\sqrt{\frac{S^2}{N}}, \overline{X} + t_{\alpha/2}\sqrt{\frac{S^2}{N}} \right] \tag{9.1}$$

as the $(1 - \alpha)$ confidence interval for μ.

What we will show here is that *the values in this interval are just those possible values of the ensemble mean of variable X which would not be rejected by a test of size α, given the observations in our data set.* Consider that a test of H_0: $\mu = \mu_0$ versus H_R: $\mu \neq \mu_0$, with size equal α, will reject H_0 if, and only if,

$$\left(\overline{X} - \mu_0 \right)\Big/\sqrt{S^2/N} \leq -t_{\alpha/2}$$

or

$$\left(\overline{X} - \mu_0 \right)\Big/\sqrt{S^2/N} > +t_{\alpha/2};$$

and it will fail to reject H_0 if, and only if,

$$-t_{\alpha/2} < \left(\overline{X} - \mu_0 \right)\Big/\sqrt{S^2/N} \leq +t_{\alpha/2}.$$

We can rewrite the last set of inequalities to bracket μ_0 and rephrase the statement to say that *the test will fail to reject $\mu = \mu_0$ under the one condition that*

$$\overline{X} - t_{\alpha/2}\sqrt{S^2/N} \leq \mu_0 < \overline{X} + t_{\alpha/2}\sqrt{S^2/N}.$$

μ_0 is not a specific value here. It is any value that might be designated by a test hypothesis. Thus we have established the claim, in this elementary context. Now we wish to establish the generic case.

Whatever the parameter θ denotes in the physical/stochastic scheme of things, our proof of the generic statement will depend on the identification of a *standardized test statistic, Z say,* on which we could base a test of hypotheses. By *standardized* we mean that *the distribution of the test statistic does not depend on the unknown parameter*. Rather, θ is contained in the formula for the statistic. In the example given above the test statistic was $Z = (\overline{X} - \mu)/\sqrt{S^2/N}$ which we knew had Student's t distribution. For the general case, we will simply say that we are able to establish the correspondences between values of p and z_p in the equation

$$p = Pr\left[Z(\theta; X_1, \ldots, X_N) \leq z_p\right]. \tag{9.2}$$

We further assume that the value of the parameter for which the test statistic is exactly equal to z_p, i.e., the inverse function

$$\theta_p = Z^{-1}(z_p; X_1, \ldots, X_N),$$

is a unique and monotonic function of p, for every possible set of observations. With these requirements we can now prove the following.

ASSERTION. *If we have constructed a $(1 - \alpha)$ confidence interval for parameter θ, from a set of observations, not only can we say that*

The probability that this interval contains

the true value of the parameter is $1 - \alpha$,

but we can also say that

This interval contains just those values of the θ

parameter which would not be rejected by a

corresponding test of size α, given

the observations in our data set.

Proof. To establish the assertion in its full generality there are three cases to consider, corresponding to "two-sided research hypotheses and to the two types of "one-sided" research hypotheses. Although it is less often of interest in a scientific context, we will establish the proof for the case in which the research hypotheses is two-sided and the null hypotheses designates a single parameter value, and leave the one-sided cases as exercises. We note that two-sided intervals and tests of hypotheses are appropriate when there are no prior constraints attached to the parameter of interest, such as a correlation which is not constrained to be either positive or negative, or a mean or variance studied under circumstances that do not dictate either an increase or a decrease. In such a situation, a $(1 - \alpha)$ confidence interval for the parameter is obtained by taking first $p = \alpha/2$ and then $p = 1 - \alpha/2$ in (9.2), so that we can write

$$1 - \alpha = Pr\left[z_{\alpha/2} < Z(\theta; X_1, \ldots, X_N) \leq z_{1-\alpha/2}\right]. \tag{9.3}$$

Here we need our assumption that the test statistic has a unique monotonic inverse. This permits us to rewrite the argument of the probability in (9.3) as a pair of inequalities that bracket θ. Then the left- and right-hand expressions give us the endpoints of the $(1 - \alpha)$ confidence interval for θ

$$Z^{-1}(z_{1-\alpha/2}; X_1, \ldots, X_N) \leq \theta < Z^{-1}(z_{\alpha/2}; X_1, \ldots, X_N). \tag{9.4}$$

Now consider testing the null hypothesis H_0: $\theta = \theta_0$ when the research hypotheses is H_R: $\theta \neq \theta_0$. For a two-sided test with size α, we calculate the value of the test statistic for $\theta = \theta_0$, and compare $Z(\theta_0; X_1, \ldots, X_N)$ with both $z_{\alpha/2}$ and $z_{1-\alpha/2}$, where the latter are derived from (9.2). When the observed value of the test statistic (with $\theta = \theta_0$) is either less than $z_{\alpha/2}$ or greater than $z_{1-\alpha/2}$, then, and only then will we conclude that the evidence is too strongly against H_0: $\theta = \theta_0$ and we reject it. But this says *exactly* that we will not reject $\theta = \theta_0$ on the one condition:

$$z_{\alpha/2} < Z(\theta_0; X_1, \ldots, X_N) \leq z_{1-\alpha/2}.$$

Comparing this with (9.3) and (9.4), we see that this is identical to θ_0 being included in the $(1 - \alpha)$ confidence interval. Since our argument used an arbitrary possible value of the parameter, then it holds for any of the possible θ values; and we have established that *the $(1 - \alpha)$ confidence interval contains just those values for θ which would not be rejected by a (two-sided) test of size α.*   ■

EXERCISES

1. (a) Write a verbal statement of a research hypothesis which is of interest to you. Translate this statement into a statement about an observable variable.
   (b) Write the antithesis of the research hypothesis: as a verbal statement and as a statement in terms of the observable variable you identified in Part (a).
   (c) Describe a data set which would be appropriate for assessing the credibility of this research hypothesis.
   (d) Name a statistic that would be appropriate for a test of these hypotheses.

2. How would you set up a test of hypotheses to resolve a dispute between two students, one from Monterey and one from Charlestown, concerning which city has more precipitation?
   (a) Specify the competing hypothesis and explain your choices.
   (b) Describe a data set which would be appropriate for resolving the issue.
   (c) Identify a statistic that could serve as test statistic. If you have some reservations about the use of this test statistic with the data set you have described, state these reservations.
   (d) Suggest a reduction of the data which would resolve the difficulty discussed in Part (c).
   Hint: Averaging values over time intervals generally reduces variability.

3. (a) For the test statistic you chose for Exercise 2 and your descriptions of H_0 and H_R, make a schematic diagram showing possible

density functions corresponding to H_0 and H_R. Chose the latter carefully, to aide in identification of a rejection region for H_0. Label these.

(b) Explain how you will find the interval endpoints for the H_0 rejection region.

(c) Could you evaluate the credibility of H_R relative to the observations, using the test statistic you chose? How or why not? What do you need to know to evaluate $cH_R ro$?

4. (a) For the set-up in Exercise 2 and with only the data from the CLIMAT file, evaluate the credibility of the null hypothesis relative to the observations ($cH_0 ro$).

(b) Write a diary note explaining the steps taken in arriving at $cH_0 ro$.

(c) What is your conclusion regarding the relative amounts of precipitation in Monterey and Charlestown? Is there any additional information you wish to have before making a statement?

5. Use precipitation data for 364 days, for Monterey and Charlestown, from CLIMAT.

(a) Make a new subfile with 52 weeks total precipitations for these two cities. Obtain time series plots versus week number, letting Week #1 be September 21 through 27.

(b) Make a new column containing the differences in the weekly total precipitations and plot this similarly.

(c) Rewrite the research and null hypotheses in terms of the weekly total precipitation differences. If necessary, redefine the test statistic and explain why you have done so.

(d) Draw a picture showing H_0 and H_R distributions, and the H_0 rejection region.

(e) Evaluate the credibility of H_0 relative to the observations.

(f) If the true variance of the weekly precipitation difference is in fact the same as the value calculated from the data, and if $\mu - \nu = 0.1$, what is the credibility of H_R relative to the data?

(g) If you used a size $\alpha = 0.15$ test for H_0 versus H_R, what would be the power of this test for $\mu - \nu = 0.1$, again assuming the calculated value of the variance of the weekly precipitation differences is the true variance?

6. Use 30-m temperature and salinity data from the STN#27 file to investigate the possible correlation between these two variables.

(a) Make a state-of-the-world decision table and fill it in for a null hypothesis of $\rho = 0$, a test of size $\alpha = 0.1$, and a research hypothesis correlation value $\rho = 0.80$.

(b) State the assumptions you have had to make in evaluating the entries of Part (a).

(c) Repeat Part (a), with $\alpha = 0.2$, again with research correlation value $\rho = 0.80$.

7. Create a subfile with monthly means of minimum temperatures and precipitation values, using data from the STCOLL file.
   (a) Choose your own criteria for the selection of data and a procedure for investigating the correlation between monthly mean MINT and PRECIP. Explain your choices.
   (b) State the research and null hypotheses; name the test statistic; choose a size for your test; and specify the set of values for which you would reject H_0.
   (c) Use MINITAB to assist you in carrying out the test of these hypotheses. Outline the process; and state your results, with your conclusion.
   (d) Evaluate the power of the test you have done, on the assumption that the true correlation is 0.4. Repeat for an assummed true correlation value of 0.7.
   (e) How many observations would you need in order to guarantee a probability of 0.65 of detecting true correlation $\rho = 0.7$?

8. (a) Compare the variances of daily maximum temperatures for May for Crookston and Saginaw, using the TEMPS file and data for 1977 through 1986.
   (b) Write a diary note explaining what you have done.
   (c) Repeat Parts (a) and (b) with data for 1950 through 1986.
   (d) How is the power of the test of hypotheses different in Parts (a) and (c)? Explain how this affects your confidence in the conclusion you reach.

9. In many research situations we have prior knowledge, or at least some firmly held expectations, that the conditions of the research program[6] will demonstrate a change in the parameter in one direction only, larger or smaller, if it demonstrates any significant change at all. In such cases, a confidence interval constructed for the parameter in question will be a one-sided confidence interval; just as the corresponding test for the research and null hypotheses will be a one-sided test. Let's say that we are only interested in the possibility that θ may be larger than some null limiting value θ_0. Prove that the $(1 - \alpha)$ confidence interval contains just those values for θ which would not be rejected by a one-sided test of size α.

10. Focus on the variance of an increment variable X which is considered to have $\mathcal{N}(0, \sigma^2)$ distribution; and assume that you have stochastically

---

[6]These "conditions" may be experimentally contrived or they may have been "contrived by the universe" and we are merely observing them.

independent observations $X_1, \ldots, X_N$. As a reference example you can think of this as the variance of Gulf Stream surface salinity measurements. Establish that the appropriate one-sided 0.90 confidence interval for the true surface salinity variance, in the region of the Gulf Stream from which you have observations, contains just those possible values σ^2 for which H_0: $\sigma^2 = \sigma_*^2$ would not be rejected with a size 0.10 test when the competing hypothesis is H_R: $\sigma^2 > \sigma_*^2$.

Hint: Find the one-sided confidence interval. Then establish the sole condition under which the test of H_0 versus H_R would not reject H_0; and show that this is equivalent to the one-sided confidence interval including σ_*^2.

10

LINEAR REGRESSION
"analyzing an influence network"

10.1 INTRODUCTION

Think of linear regression in connection with explaining variations in one *primary variable* (PV) in terms of changes in other *subsidiary variables* (SVs). This is the objective here. To achieve it we will generalize techniques we have already met and used for determining whether, and to what extent, variations in one observed quantity are associated with variations in another. Then we will develop the technology to create a mathematical representation for the primary (increment) variable in terms of subsidiary (increment) variables, with coefficients determined from data. Finally we combine distribution theory with hypothesis testing, in the structure for evaluating the significance of estimated relationships.

There are different reasons for wanting to use a linear regression analysis. One is diagnostic; for example, using a formulation of the relationship between variables as an avenue to learning more about the physical, chemical, or biological properties of a system. Another is estimating or predicting the primary variable in circumstances in which it is easier to observe values of the subsidiary variables than of the primary variable. In fact, sometimes "easier" will mean that *the subsidiary variables can be observed, whereas the primary variable cannot be.* The following provide examples.

1. Next month's precipitation amount, when we have a precipitation history;
2. Ten-meter ocean temperatures along a trawl line for which we know surface temperatures and wind stress;
3. 700 mb winds over the Gulf of California, with surface and 850 mb wind values augmented by wind and geopotential soundings from radiosondes east of the Gulf.

Each of these examples describes a situation in which data for all relevant variables, including the PV, can be collected in a field experiment and put into a data base. With this we can parameterize a mathematical description of the relationship of the primary variable to the subsidiary variables. This mathematical description is the central objective of regression analysis, and forms the basis for diagnostics or for estimation in the future when we may not have values of the PV. *Linear regression assumes such a data base is available.* Linear regression also assumes that the mathematical description of the relationship is a linear combination of subsidiary variables. The presumed linearity of the combination is not as restrictive as its first appearance may suggest. It may be possible to transform initial subsidiary variables, whose relationships with the PV are demonstrably nonlinear, prior to their addition to a linear estimator of the PV. This simply redefines the set of subsidiary variables, and gives us new SVs for which the presumption of linearity of their relationships with the primary variable has greater validity. In what follows we assume that any required transformations have been made in creating the data base, so that we can deal in purely linear relationships.

Usually it is not a "field variable" for which we develop a predictor algorithm, but the *difference* between the value that might be observed and a concurrent, known "guess value". We call this difference an *increment*. Specifically, increments are those portions of variables as they are (or would be) observed, that are not accounted for by concurrent guess values. The "guess values" are generally estimates of ensemble mean values which have been computed from the current or a prior data base; although in some applications increments may be as sophisticated as the differences between observed (or observable) values and corresponding output from complex physics-based numerical models. Our reason for making such a point of the distinction between observed (and observable) variables and increments defined from them, vis-à-vis linear regression, is to draw attention to the fact that this analysis technique does not presume that field variables are themselves linearly related. This point is critical to the usefulness of the techniques presented in this chapter. *The hypothesized linearity applies to the increments.* In a later section we will relax even this assumption, and offer a guide to techniques for identification and management of demonstrable nonlinearities. These techniques give us a powerful capability for carrying out sophisticated analyses. As noted, when the increments of primary and subsidiary variables have a relationship which is clearly nonlinear, a transformation of one or the other may lead to a more nearly linear relationship. In this connection, the graphics capabilities of standard software packages become invaluable aids to interactive "data sleuthing" for the detection and representation of simple but nonlinear relationships.

Regression analysis focuses on explaining residual variation by analyzing discrepancies between what we have observed and a synopsis of what we anticipate from a hypothesized relationship between the subject variables. "What we anticipate" is specified by the mathematical description of the relationship of the primary variable to the subsidiary variables when values have been given to the coefficients of the linear expression. The mechanism for selecting values for the coefficients invokes a strong optimality criterion which has both intuitive appeal and the stature of an old and respected mathematical derivation (dating from K. F. Gauss' work in 1795). This criterion is predicated on characteristics of the joint distributions of the stochastically dependent, increment variables. The assumed normality of the joint distribution and utilization of the observed variances and covariances of the increments are preeminent in the structure and application of diagnostic and prediction procedures of regression analysis.

In the following, we establish the conceptual and computational relationship between the minimum-squared-error criterion for a regression estimate of a primary variable increment and the observed variances and covariances of primary and subsidiary increment variables. Once we have developed structure for the regression representation and acknowledged the assumptions on which its utilization rests, we can call on standard software for applications. In two- and multi-variable regressions, we will show how this software may be used to parameterize estimators and evaluate the significance of conjectured relationships.

10.2 TWO-VARIABLE LINEAR REGRESSION

Two-variable linear regression is simple to handle notationally; and we can treat it is a direct extension of the analysis of correlation between two variables, which we discussed earlier. (You may wish to reread the last two paragraphs of Section 8.5 and all of Section 8.6, to refresh your memory.) Analysis of two-variable linear regression differs from analysis of correlation in that the regression objective is to find a formula for describing one variable in terms of the other. In addressing the correlation between two variables, per se, there was no delegation of preeminence to one of the variables, only an evaluation of the strength of a linear relationship between their increments. Here a natural precedence is defined by the analysis objective, because the linear relationship is the basis for prediction or estimation of the primary variable as a function of the subsidiary variable. Although it is generally inappropriate in ocean and atmospheric science to think of variability in the one as being in response to variability on the other, classical statistical terminology labels the two variables as "dependent" and "independent", respectively. We will refer this concept of dependence strictly to the relationship between the mathematical representations of their increments.

We will denote the primary and subsidiary (increment) variables by Y and X, and write our objective as the construction of an estimator for Y as a linear function of X,

$$\hat{Y} = a + bX. \tag{10.1}$$

a is the intercept value for $\hat{Y}$ at $X = 0$, and b is the slope of the regression line that gives us values for $\hat{Y}$ from values observed for X. We make and note the assumption that this linear relationship between subsidiary variable X and primary variable Y is the same in the prescribed data base as it will be in those circumstances in which we will employ (10.1) to estimate Y. Subject to the truth of this assumption, the data base can be used to evaluate the predictor coefficients a and b. The optimality criterion that we apply for the derivation of their optimal values is minimization of the sum of squared deviations,

$$SS = \sum_{j=1}^{N} \left(Y_j - \hat{Y}_j\right)^2 = \sum_{j=1}^{N} \left[Y_j - \left(a + bX_j\right)\right]^2, \tag{10.2}$$

where the sum is over the paired observations of Y and X. Specifically this formulation estimates each observed value Y_j with corresponding $a + bX_j$ and computes the total of the squared errors of these estimates: The "deviations" in SS are the differences between Y_j observations and what is estimated from X_j. To find a and b values that minimize SS, we obtain the derivatives of the sum of squares with respect to a and b, and solve simultaneously for the values for which the derivatives equal zero[1]

$$0 = \left.\frac{\partial SS}{\partial a}\right|_{(a,b)=(\hat{a},\hat{b})} = \left.-2\sum_{j}\left[Y_j - \left(a + bX_j\right)\right]\right|_{(a,b)=(\hat{a},\hat{b})}$$

$$0 = \left.\frac{\partial SS}{\partial b}\right|_{(a,b)=(\hat{a},\hat{b})} = \left.-2\sum_{j}X_j\left[Y_j - \left(a + bX_j\right)\right]\right|_{(a,b)=(\hat{a},\hat{b})}.$$

[1]Technically all we can say is that the values that make first derivatives zero give us an extreme point of the sum of squares: either a minimum or a maximum. However, you can confirm that $SS(\hat{a},\hat{b})$ is the minimum point, by finding second derivatives and discovering that they are both positive at $(\hat{a},\hat{b})$, thus guaranteeing a minimum.

These give us

$$\hat{a} = \bar{y} - \hat{b}\bar{X} \quad \text{and} \quad \hat{b} = \sum_j \left(Y_j - \bar{Y}\right)\left(X_j - \bar{X}\right) \Big/ \sum_j \left(X_j - \bar{X}\right)^2.$$

The latter may be written expressly in terms of the observed covariance and the variance of X, as

$$\hat{b} = S_{YX}/S_{X^2}.$$

In fact, since the observed correlation between Y and X for paired observations $(Y_1, X_1), \ldots, (Y_N, X_N)$ is

$$R = \sum_j \left(Y_j - \bar{Y}\right)\left(X_j - \bar{X}\right) \Big/ \sqrt{\sum_j \left(Y_j - \bar{Y}\right)^2 \sum_j \left(X_j - \bar{X}\right)^2} = S_{YX}\Big/\sqrt{S_Y^2 S_X^2}$$

we may also write the estimated slope of the regression line as

$$\hat{b} = R\sqrt{S_Y^2/S_X^2}.$$

Thus we see that, given constant variances, our best estimate of the rate of change of the primary variable with the subsidiary variable is proportional to their observed correlation.

Our linear regression objective is obtained by writing a and b into (10.1) as

$$\hat{Y} = \left(\bar{Y} - \hat{b}\bar{X}\right) + \hat{b}X = \bar{Y} + \hat{b}(X - \bar{X})$$

$$= \bar{Y} + (X - \bar{X})S_{YX}/S_X^2 \tag{10.3}$$

or

$$\hat{Y} = \bar{Y} + (X - \bar{X})R\sqrt{S_Y^2/S_X^2}.$$

In (10.3) we have dropped the subscript j that designated a pair of values in the data set. We have written the regression expression as we will use it: as a general diagnostic and estimator. It will serve for any pair (Y, X) to describe the relationship between these variables generically, and permit us to estimate Y when we can observe only X.

Another way of thinking about the total of the squared errors (10.2) is that this is the *residual sum of squares*, namely, *what is left of the measure of the total variability of the primary variable after we have made optimal use of the information in the data to account for its linear relationship to the subsidiary variable.* We may substantiate this claim mathematically by partitioning the total sum of squares of the primary variable around its own observed mean as shown by

$$SS_{total} = \sum_{j=1}^{N} \left(Y_j - \bar{Y} \right)^2 = \sum_{j=1}^{N} \left[\left(Y_j - \hat{Y}_j \right) + \left(\hat{Y}_j - \bar{Y} \right) \right]^2$$

$$= \sum_{j} \left(Y_j - \hat{Y}_j \right)^2 + \sum_{j} \left(\hat{Y}_j - \bar{Y} \right)^2$$

$$= SS_{res} + SS_{reg}. \tag{10.4}$$

As you can confirm with a little algebra, the cross product term in the expansion of the sum of squares on the first line is zero. We have not taken the space to show this, but leave it as an exercise.

The first sum of squares on the right in the middle line of (10.4) is the *residual sum of squares*. This is (10.2) minimized by our choice of $\hat{a}$ and $\hat{b}$; and we label it SS_{res}. The second sum of squares is the difference between the total sum of squared deviations of the Y's from their mean, and the residual sum of squares, i.e.,

$$\sum_{j} \left(\hat{Y}_j - \bar{Y} \right)^2 = \sum_{j} \left(Y_j - \bar{Y} \right)^2 - \sum_{j} \left(Y_j - \hat{Y}_j \right)^2.$$

Thus it is that part of the total variation of the observed values of the primary variable Y which has been *explained* by the regression on subsidiary variable X; and we label this term SS_{exp}. Some statistics books denote the residual sum of squares by SS_{error}, a convention that identifies it as the portion of the total sum of squares that estimates error in the observations of Y. Also, in some references you may find the explained sum of squares denoted by SS_{reg}, for *sum of squares due to regression*.

We can view the elements of SS_{total} and SS_{res} in graphical form as shown in Figs. 10.1(a) and 10.1(b). In Fig. 10.1(a), the lengths of the dotted lines are the differences between the individual Y_j's and their observed mean $\bar{Y}$. These are the differences which are squared and summed in SS_{total}. In Fig. 10.1(b), the increments are the differences in SS_{res}, namely, the differences between the observed Y_j's and the values predicted by the regression line when we evaluate it at the paired X_j's. In this illustration, there remains considerable residual variation of the Y_j values, even around the best regression line. However, it is evident, and it will always be the case, that the overall deviation of the Y's from the line that represents their trend with X is smaller than their deviation from their mean $\bar{Y}$. Determining whether the regression is "significant", i.e., whether there is enough evidence to support reporting and defending a linear expression for the relationship between primary and subsidiary variables, is one of the tasks regression analysis is designed to fulfill. We note that $\hat{Y} = \bar{Y}$ is a special case of a regression of Y on X, namely the null case, for

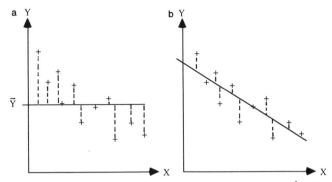

FIGURE 10.1 (a) Plot of observed (Y_j, X_j) increments with the line $\hat{Y} = \bar{Y}$ and vertical dotted segments showing deviations of Y_j values from $\bar{Y}$. (b) Plot of observed pairs (Y_j, X_j) with the line $\hat{Y} = \hat{a} + \hat{b}X$ and vertical dotted segments showing deviations of Y_j values from the regression line.

which $b = 0$. Thus the null hypothesis for any research hypothesis that specifies a relationship between Y and X may be paraphrased as a regression line with slope zero; and a test of hypotheses for regression is a determination of whether or not we can conclude that the slope is non-zero.

When we use MINITAB to carry out the computations for an analysis of the regression of Y on X, the output presents us with a table. This displays the bookkeeping for the analysis in standard format; and it is called an analysis of variance (ANOVA) table. The following example includes an illustration.

Example of Two-Variable Linear Regression

Recall that the *median* of a set of numbers is a value that has as many numbers of the set that are greater as there are smaller. Thus it is the *middle value in the frequency sense.* Here we use January temperature data from State College, for the 10 years of the 1980s, and create two statistics for each year: the median of the 31 daily maximum values (MEDMAX) and the median of the 31 daily minimum values (MEDMIN). These statistics are displayed in Table 10.1; and Fig. 10.2 shows a plot of the 10 pairs of values. As we would expect, there is evidence of a linear relationship between these variables (although perhaps it is not as strong as we might have anticipated).

With the pair of values $Y = $ MEDMAX and $X = $ MEDMIN characterizing each January of the 10 years, we will determine how much of the variability of the Y values in our short record can be described by a linear

TABLE 10.1   January Temperature Summary Values

Year	MEDMAX	MEDMIN
1980	34° F	19° F
1981	28	13
1982	25	12
1983	35	23
1984	32	14
1985	28	15
1986	35	19
1987	33	23
1988	30	8
1989	40	21

relationship between Y and X. Since the data record is short, it would not be arduous to do this "long hand", first using the data in the table to obtain $\hat{a}$ and $\hat{b}$, so that we can write the predictor expression $\hat{Y} = \hat{a} + \hat{b}X$, and then calculating SS_{total} and SS_{reg}, for comparison. However MINITAB will respond to one command with all the calculated values, if we enter the data of Table 10.1 in two columns of a MINITAB worksheet, label them 'MEDMAX' and 'MEDMIN', and use

REGRESS 'MEDMAX' on 1 predictor 'MEDMIN'.

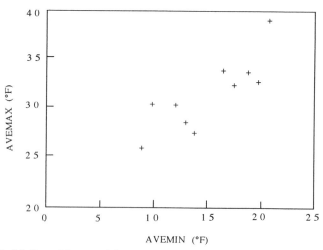

FIGURE 10.2 Median of January daily maximum temperatures versus median of January daily minimum temperatures, for the years 1980–1989, obtained from the temperature records for State College, PA.

The output will look like the following:

THE REGRESSION EQUATION IS
MEDMAX = 21.6 + 0.6 MEDMIN

PREDICTOR	COEF	STDEV	T-RATIO	P
CONSTANT	21.6	3.7	5.9	0.00
MEDMIN	0.6	0.2	2.9	0.02

S = 3.2   R-SQ = 52%   R-SQ(adj) = 46%

ANALYSIS OF VARIANCE

SOURCE	DF	SS	MS	F	P
REGRESSION	1	89	89.0	8.6	0.02
ERROR	8	83	10.4		
TOTAL	9	172	19.1		

There will be additional information in the computer output which we have not displayed. MINITAB provides it to assist you in looking at fine points of the analysis, such as identifying observations that are unusual in the sense that the regression fit to them may be particularly bad. These values will have correspondingly large residuals and, consequently, outsized influences on analysis results and their possible interpretation. Thus it is particularly important to confirm the validity of these observations. The additional output can be especially helpful if you have a large data set for which it is difficult or impractical to view each value in the context of the full set. As a by-product of the analysis, it will draw your attention both to observations that may contain errors not caught in data processing, as well as to unsuspected but scientifically interesting observing situations. This output is generally worth scanning with these possibilities in mind.

What we get immediately from the MINITAB output shown above, that we are in a position to discuss at this point in our development of the subject of regression analysis, are the values for the coefficients permitting us to write

$$\hat{Y} = 21.1 + 0.52(X - \bar{X})$$

and the values for both total and residual sums of squares

$$SS_{total} = 172 \quad \text{and} \quad SS_{res} = 83.$$

Consequently the information provided by X in its *linear* relationship to Y has "explained" almost half the variability of the latter, in this example.

We will interpret more of the items in the output when we have considered more aspects of regression analysis. We will do this in the context of regression with several subsidiary variables.

10.3   LINEAR REGRESSION WITH SEVERAL VARIABLES

Linear regression with several variables is a generalization of two-variable linear regression to a situation that we encounter frequently in research. Here we address the twin challenges of estimating influences and accounting for variability when there are several subsidiary variables, known or assumed to be closely associated with the primary variable. The subsidiary variables may be correlated with one another. Undoubtedly at least some will be, and the analysis will account for their correlation. However, we exclude the possibility that any of the subsidiary variables are redundant, and assume that each is considered for its potential and unique influence on the primary variable.

Again our objective is the construction of an estimator for Y. Here the estimator is a linear expression in the several subsidiary variables

$$\hat{Y} = a + b_1 X_1 + \cdots + b_K X_K, \qquad (10.5)$$

where a is the intercept value for $\hat{Y}$ at the point where all K subsidiary (increment) variables are zero. The "slope" interpretation of each of the remaining regression coefficients is not as useful here as it was in the case of two-variable regression; and we will think of the values of the b's simply as relative weights. Again we seek first to estimate the regression coefficients from data and then to use the estimates in inferences about the variability of the primary variable, Y. To accomplish both objectives we must make the assumption that, whatever the true linear relationship is, it is the same in those circumstances for which we have observations as it is in the circumstances for which we cannot observe Y but wish to estimate it from observations of $X_1, \ldots, X_K$.

The Optimal Estimator for Y

We begin the estimation discussion by simple extension of the single subsidiary variable case. Then, anticipating the next section in which we look at the theoretical basis for regression inferences, we write the objective and its solution in matrix notation. The criterion applied in deriving optimal estimates for the regression coefficients is minimization of the sum of squared deviations analogous to (10.2),

$$SS = \sum_{j=1}^{N} \left[Y_j - \left(a + b_1 X_{1j} + \cdots + b_K X_{Kj} \right) \right]^2, \qquad (10.6)$$

where the sum is over the full data base, assumed to have N simultaneous observations of Y and the K subsidiary variables. Again the formulation assumes that we estimate each observed value Y with corresponding $a + b_1 X_{1j} + \cdots + b_K X_{Kj}$ and compute the total of the squared errors of the estimates. In this multiple regression case, we will find $\hat{a}$ first, in terms of $\hat{b}_1, \ldots, \hat{b}_K$ (whatever they turn out to be) and then find the $\hat{b}$ values. Using

$$0 = \frac{\partial SS}{\partial a}\bigg|_{(a, b_1, \ldots, b_K) = (\hat{a}, \hat{b}_1, \ldots, \hat{b}_K)}$$

$$= -2\sum_j \left[Y_j - (a + b_1 X_{1j} + \cdots + b_K X_{Kj}) \right]\bigg|_{(a, b_1, \ldots, b_K) = (\hat{a}, \hat{b}_1, \ldots, \hat{b}_K)}$$

we find

$$\hat{a} = \bar{Y} - \sum_k b_k \bar{X}_k. \tag{10.7}$$

With this we can simplify minimization of the sum of squared errors for the linear regression estimate of Y, with respect to the b coefficients by first rewriting (10.6) as

$$SS = \sum_j \left(Y_j - \hat{Y}_j \right)^2 = \sum_j \left[\left(Y_j - \bar{Y} \right) - \sum_k b_k \left(X_{kj} - \bar{X}_k \right) \right]^2.$$

In these equations we have substituted $\hat{a}$ from (10.7) and regrouped the terms into differences between the observations for each variable and their corresponding observed means. Using the prime notation (') to indicate that the corresponding mean has been subtracted from each value

$$Y'_j = Y_j - \bar{Y} \quad \text{and} \quad X'_{kj} = X_{k_j} - \bar{X}_k, \quad \text{for } k = 1, \ldots, K,$$

for each set of observations $j = 1, \ldots, N$, we can now write (10.5) in matrix array notation as

$$\begin{pmatrix} Y_1 - \bar{Y} \\ \vdots \\ Y_N - \bar{Y} \end{pmatrix} = \begin{pmatrix} X_{11} - \bar{X}_1 & \cdots & X_{K1} - \bar{X}_K \\ \vdots & & \vdots \\ X_{N1} - \bar{X}_1 & \cdots & X_{NK} - \bar{X}_K \end{pmatrix} \begin{pmatrix} b_1 \\ \vdots \\ b_K \end{pmatrix} \quad \text{or} \quad \mathbf{Y}' = \mathbf{X}'\mathbf{b}$$

and the sum of squared errors of (10.6) as

$$SS = \left(\mathbf{Y}' - \hat{\mathbf{Y}}' \right)^T \left(\mathbf{Y}' - \hat{\mathbf{Y}}' \right) = \left(\mathbf{Y}' - \mathbf{X}'\mathbf{b} \right)^T \left(\mathbf{Y}' - \mathbf{X}'\mathbf{b} \right). \tag{10.8}$$

In the context of a given data set, this as a function only of the regression coefficient vector $\mathbf{b}$. Equation (10.8) is a cumulative error function with its minimum value at

$$\hat{\mathbf{b}} = \left(\mathbf{X}'^T \mathbf{X}' \right)^{-1} \mathbf{X}'^T \mathbf{Y}'. \tag{10.9}$$

For a proof that (10.9) does indeed minimize (10.8) see Exercise 10.8.

Now we can write the optimal estimator for Y in terms of the data set, using the coefficients $\hat{a}$ and $\hat{\mathbf{b}}$ found to minimize the total sum of squared errors, and the universal unit vector $\mathbf{e}^T = (1 \ldots 1)$:

$$\hat{Y} = \hat{a}\mathbf{e} + \mathbf{X}'\hat{\mathbf{b}} = \bar{Y}\mathbf{e} + (X'^T X')^{-1} X'^T Y'. \qquad (10.10)$$

Partitioning the Total Sum of Squares

Just as we found with a single subsidiary variable X, the total of the squared deviations of the observed values of the primary variable from its observed mean may be written as the sum of two component sums of squares. One contains all of the observed variability of Y that cannot be associated with a linear function of the X's, and the other is the sum of squared deviations of the regression expression for Y from the observed mean. We can write the multiple regression analogue to the partitioning of the sum of squares (10.4) as

$$\begin{aligned} SS_{\text{total}} = \mathbf{Y}'^T \mathbf{Y}' &= \left[(\mathbf{Y} - \hat{\mathbf{Y}}) + (\hat{\mathbf{Y}} - \bar{Y}\mathbf{e}) \right]^T \left[(\mathbf{Y} - \hat{\mathbf{Y}}) + (\hat{\mathbf{Y}} - \bar{Y}\mathbf{e}) \right] \\ &= (\mathbf{Y} - \hat{\mathbf{Y}})^T (\mathbf{Y} - \hat{\mathbf{Y}}) + (\hat{\mathbf{Y}} - \bar{Y}\mathbf{e})^T (\hat{\mathbf{Y}} - \bar{Y}\mathbf{e}) \\ &= SS_{\text{res}} + SS_{\text{reg}}. \end{aligned} \qquad (10.11)$$

Since SS_{res} contains the "residual squared deviations" of the observed values of Y that remain after an optimal accounting for the variability associated with $\mathbf{X}$, we can use this sum of squares to estimate the variance of the errors in the observations. In addition we will establish that SS_{reg} provides an independent estimate of this error variance *plus* a function of the estimated linear relationship between Y and $\mathbf{X}$. We can use them together to test hypotheses about the true relationship between Y and $\mathbf{X}$. To this end we will need the degrees of freedom associated with each component sum of squares.

Corresponding to the partitioning of the total sum of squared deviations, there is a partitioning of the associated degrees of freedom:

$$DF_{\text{total}} = DF_{\text{res}} + DF_{\text{reg}}.$$

The total degrees of freedom, on the left, is $N - 1$ since we have used the N sources of information contributing to the total sum of squares $\mathbf{Y}'^T \mathbf{Y}'$ to estimate one parameter, and thus lost one degree of freedom from the original N. The first term on the right is the number of degrees of freedom associated with $(\mathbf{Y} - \hat{\mathbf{Y}})^T (\mathbf{Y} - \hat{\mathbf{Y}})$ which may also be written in terms of the mean deviations as $(\mathbf{Y}' - \hat{\mathbf{Y}}')^T (\mathbf{Y}' - \hat{\mathbf{Y}}')$. To evaluate this expression from any data set, we must first use the N original sources of information to estimate the K parameters $b_1, \ldots, b_K$, in addition to estimating the mean. Thus, conceptually, there are $K + 1$ constraints on the number of sources of variability contributing to the residual sum of squares and its associated degrees of freedom must be reduced by this amount, that is, $DF_{res} = N - (K + 1) = N - K - 1$. Finally, we get the degrees of freedom for the sum of squares due to regression as their difference:

$$DF_{reg} = DF_{total} - DF_{res} = (N - 1) - (N - K - 1) = K.$$

Example of Linear Regression with Several Variables

For this illustration we will use the data set shown in Table 10.2, containing the daily observations of seven variables, for State College, for the month of October 1990. Clearly we cannot plot the data in the table in any

TABLE 10.2   October 1990 Observations for State College, PA

Date	Pressure	Ave Temp	T	DPT	RH	Wnd Spd	Peak Wnd
1	30.0	55.4	19	39.3	56	5.6	25
2	0.2	51.0	24	35.2	57	4.1	19
3	0.2	60.1	26	44.4	57	5.6	17
4	29.9	56.9	13	49.7	77	6.1	16
5	30.1	63.7	17	44.1	50	5.2	19
6	0.1	66.7	28	46.0	48	4.1	16
7	0.2	67.0	25	50.5	55	2.2	12
8	0.1	68.1	17	57.5	68	1.0	10
9	0.4	70.2	16	60.5	71	3.2	13
10[a]							
11	0.1	56.6	15	53.0	88	1.6	18
12	0.0	65.1	17	59.4	81	1.2	12
13	29.8	65.1	15	57.8	77	1.6	8
14	0.9	58.2	62	46.6	66	2.5	13
15	30.1	52.3	22	35.4	55	4.7	17
16	0.3	53.9	20	36.2	53	1.8	9
17	0.1	63.7	8	48.1	57	5.6	18
18	29.7	45.6	26	34.5	67	12.2	31
19	30.2	39.1	19	27.8	67	4.5	21
20	0.3	47.7	26	31.2	56	2.4	10
21	0.2	53.2	17	42.2	68	2.9	12
22	0.0	54.3	9	50.9	89	0.2	6
23	29.8	54.6	10	50.1	85	1.9	14
24	0.9	48.1	55	36.9	67	2.2	11
25	0.9	46.6	20	34.0	64	3.9	16
26	30.1	34.6	17	20.1	60	3.3	17
27	0.1	44.7	25	26.0	51	3.5	18
28	0.1	38.9	13	26.0	63	9.0	21
29	0.3	38.4	21	22.4	56	2.5	13
30	0.2	54.8	32	28.6	39	3.6	12
31	0.2	53.6	20	37.0	55	0.2	5

[a]The line in the table showing data for October 10 is blank because the data are missing for that one day.

way that will show the relationships among all of the variables, simultaneously. However, we can use MINITAB to explore their ensemble relationships in other ways. With the data in a MINITAB worksheet, in columns as they are shown here, we can obtain the pairwise correlations for as many of the variables as we wish, with a single command. For example, in response to

"CORRELATION matrix for C2-C7"

MINITAB prints out a table

	C2	C3	C4	C5	C6	C7
PRESSURE	1.00					
AVE TEMP	0.01	1.00				
DELT	−0.13	−0.06	1.00			
DPT	−0.21	0.88	−0.18	1.00		
RH	−0.46	0.11	−0.28	0.56	1.00	
WND SPD	−0.26	−0.27	−0.04	−0.32	−0.23	1.00

If we focus on RH, which is relative humidity, we see that its relationships with the non-humidity variables are ordered by their correlation magnitudes as

PRESSURE: $R = -0.46$, DELT: $R = -0.28$,
WND SPD: $R = -0.23$, AVE TEMP: $R = 0.11$.

We can put them all into a regression expression for RH, together. Or we can begin with the most highly correlated of the subsidiary variables, in a two-variable linear regression, and then add one at a time, in order to study the progression of refinement of the accounting for the variability of the primary variable. Here we choose to do the latter, because we recommend it as a method of exploratory data analysis.

We will use the R^2 *statistic*, which is *the fraction of total variation in the primary variable explained by the regression*, to study the progression of results as we increase the number of subsidiary variables included. This statistic is one of the elements in the output of a MINITAB regression analysis where it appears as R-SQ. It is the ratio of the residual sum of squares to the total sum of squares for the primary variable

$$R^2 = \frac{\sum\limits_{j=1}^{N} \left(Y_j - \hat{Y}_j\right)^2}{\sum\limits_{j=1}^{N} \left(Y_j - \overline{Y}\right)^2}.$$

For the present example we will not reproduce the MINITAB output in any detail. Instead we present a summary of the results, giving the

TABLE 10.3 Multivariate Regression Results for RH

MINITAB Command	$R^2 \times 100$
REGRESS C6 1 C2	20.9%
REGRESS C6 2 C2, C4	32.7%
REGRESS C6 3 C2, C4, C7	47.7%
REGRESS C6 4 C2, C3, C4, C7	47.8%

successive regression commands together with the output R-SQ measures of explained regression, as we add variables one at a time, in the order of their correlations with the relative humidity variable. These are shown in Table 10.3. In the table the designation of the primary variable in the command is its column label, here C6 for the values of relative humidity. The integer following C6 is the number of subsidiary variables in the regression equation, increasing from 1 to 4 in successive commands. This is followed in turn by the column designations for the variables to be included.

In the summary of output we can see that adding additional variables significantly increases the fraction of total variation explained by the regression, over that of the single most highly correlated subsidiary variable. However, the final addition, including C3 (AVE TEMP) in a regression equation with C2, C4, and C7, brings only a small increase in the fraction of explained variability. We might have anticipated this from an examination of the correlation table which shows the correlation between RH and AVE TEMP to be 0.11. Nonetheless, going through this step-by-step process of addition to the regression equation highlights an important point: namely, that it may not be wise to include all available data indiscriminately. Relationships among the subsidiary variables themselves, that are of no direct interest, may detract from the clarity of interpretation of the output. It is well to heed advice of one of the modern giants of statistics who stressed the importance of parsimony in the specification of statistical models. We will return to this example when we have established the basis for testing regression hypotheses, to illustrate this final point.

10.4 THE BACKBONE OF REGRESSION ANALYSIS IN STATISTICAL THEORY

So far our discussion of the relationship between Y and X has not been in the context of the stochastic distribution of (Y, X). This context provides the statistical backbone of regression analysis, which we will need to assess evidence for the validity of a hypothesized relationship, to find confidence intervals for the estimated regression coefficients, and to put confidence bounds on predicted values of the primary variable for values of X for which we cannot observe Y.

To achieve the generality we require for multiple linear regression we address the more general case in which the subsidiary variable $\mathbf{X}$ is vector valued. Then we will have all our bases covered with one notation.

In the largest most general framework for the statistical analysis of several variables at a time, we would start with the assumption that the observations available to us are observed values of $K+1$ variables which have a joint (multivariate) distribution. If we could then argue that it is reasonable to assume that the joint distribution is Normal, we could go on to establish that the form of *the best estimator for Y in terms of the components of* $\mathbf{X}$ *is a linear combination of* $X_1, \ldots, X_K$. (See Section 4.1 of Thiébaux and Pedder, 1987, which also establishes the coefficient array for the optimal estimator as a known function of the covariance structure of the $(K+1)$-dimensional distribution.) When we refer to this as "the best, or optimal estimator" we mean that it is best in the sense that, of *all* the functions that might be selected to estimate the primary variable Y, the *one* that will have the smallest ensemble squared difference from the true values of Y is a linear expression

$$\hat{Y} = a + b_1 X_1 + \cdots + b_K X_K.$$

This result is the primary rationale for the widespread use of the linear regression formulation.

We will not go further with this chain of basic assumptions about joint distributions, and their consequences, because the analysis technology is highly dependent on the form of the joint distributions, through their moment structures. Generally we will not know this. Neither are we likely to have "outsider knowledge" of the covariance functions that govern a Normal system, even though we may have valid arguments for believing that the joint distribution of Y and $\mathbf{X}$ is Multivariate Normal. There are many excellent references, for readers who wish to learn more than we present here about the rich and multifaceted mathematical science of analyzing collective stochastic behavior of several variables. A few of the classic references are Brownlee (1965), Draper and Smith (1981), Graybill (1961), and Scheffé (1959). To do justice to this vast subject, in the present volume, would require many additional chapters and detract from our central purpose here.

We believe it is safe to say that in most research settings that we are likely to encounter in atmosphere and ocean sciences, our knowledge of characteristics of the covariate stochastic behavior of the (increment) variables of our study is limited to what we can learn from the present data set. Thus our descriptions of these characteristics are derived from the same data we use to make inferences about the relationships among the variables.[2] The way in which we will deal with this situation is to explicitly qualify the inferences we make about Y and its relationships with the X's, as dependent on the present data set. Specifically, we take the observed values of $X_1, \ldots, X_K$ as *given*, and *given these values* we hypothesize

$$Y = a + b_1 X_1 + \cdots + b_K Y_K + \varepsilon, \qquad (10.12)$$

where the only stochastic element on the right-hand side is ε. The expression (10.12) says that the stochastic component comprises the full discrepancy between whatever Y value is observed and the hypothesized linear combination of the given X's. Here and throughout the remainder of the chapter, we will not only assume that ε has a Normal distribution with ensemble mean equal to 0 and a variance that is the same for every situation for which we have observations, but that these discrepancies are stochastically independent of one another, for different situations.

We can rewrite (10.12) to include a full set of observations in a single vector notation, and our assumptions about the stochastic component, as

$$\underset{N \times 1}{\mathbf{Y}} = \underset{N \times 1}{a\mathbf{e}} + \underset{N \times K}{\mathbf{X}} \underset{K \times 1}{\mathbf{b}} + \underset{N \times 1}{\boldsymbol{\varepsilon}}, \qquad (10.13)$$

where

$$\boldsymbol{\varepsilon} \to \mathcal{N}(\mathbf{0}, \sigma^2 I).$$

The notations beneath Eq. (10.13) identify the dimensions of the corresponding arrays. You may note that we have again chosen to write vector arrays as columns vectors, i.e., as $N \times 1$ or $K \times 1$ matrices, and adopted the unit vector and identity matrix notations: $\mathbf{e}$ to denote a column vector of 1's and I to denote a square matrix, with diagonal entries equal to 1 and 0's elsewhere.

Since we regard the $\mathbf{X}$ as given, the assumptions we have made about $\boldsymbol{\varepsilon}$ imply that

$$\mathbf{Y} \to \mathcal{N}(a\mathbf{e} + \mathbf{Xb}, \sigma^2 I). \qquad (10.14)$$

Ensemble Means of the Sums of Squares and Mean Squares

The components of the total sum of squared deviations of the primary variable from its observed mean are central to analysis of the evidence in the data set bearing on a hypothesized relationship between the primary and subsidiary variables, as suggested at the end of the last section. Here

---

[2] To estimate the characteristics of an assumed distribution and use them as "ground truth" for scientific inferences made with statistics calculated from the same data set would seriously compromise the validity of the inferences, and thus undermine our scientific objective.

we put this on more definite footing by establishing the statistical expectations (ensemble means) of the residual and regression sums of squares of the partitioning (10.11), with the expected values of the sums of squares derived from the distribution of the stochastic element $\boldsymbol{\varepsilon}$.

ASSERTION   10.1. $\mathscr{E}[(\mathbf{Y} - \bar{Y}\mathbf{e})^T(\mathbf{Y} - \bar{Y}\mathbf{e})] = (N - 1)\sigma^2 + \mathbf{b}^T(\mathbf{X} - \mathbf{e}\bar{\mathbf{X}})^T(\mathbf{X} - \mathbf{e}\bar{\mathbf{X}})\mathbf{b}$

and

$$\mathscr{E}\left[\left(\mathbf{Y} - \hat{\mathbf{Y}}\right)^T\left(\mathbf{Y} - \hat{\mathbf{Y}}\right)\right] = (N - K - 1)\sigma^2.$$

Proof. Part (a).   In order to state and prove our claims for these values of SS_{total} and SS_{res} we will recall some notational conventions and then put them to use to establish simple consequences of (10.13). Specifically, we can write arithematic means as

$$\bar{Y} = \mathbf{e}^T\mathbf{Y}/N, \qquad \bar{\mathbf{X}} = \mathbf{e}^T\mathbf{X}/N = \left(\bar{X}_1 \cdots \bar{X}_K\right), \qquad \bar{\varepsilon} = \mathbf{e}^T\boldsymbol{\varepsilon}/N$$

and apply these to (10.13) to get

$$(\mathbf{Y} - \bar{Y}\mathbf{e}) = (\mathbf{X} - \mathbf{e}\bar{\mathbf{X}})\mathbf{b} + (\boldsymbol{\varepsilon} - \mathbf{e}\bar{\varepsilon}). \tag{10.15}$$

If we define $\boldsymbol{\zeta}$ as the variable in the last bracketed expression and write it as

$$\boldsymbol{\zeta} = (\boldsymbol{\varepsilon} - \mathbf{e}\bar{\varepsilon}) = (\mathbf{I} - \mathbf{e}\mathbf{e}^T/N)\boldsymbol{\varepsilon},$$

we recognize it as a simple linear transformation of $\boldsymbol{\varepsilon} \to \mathscr{N}(\mathbf{0}, \mathbf{I})$. Now we can apply Assertion 7.4 to get

$$\boldsymbol{\zeta} \to \mathscr{N}(\mathbf{0}, \boldsymbol{\Lambda}),$$

where

$$\boldsymbol{\Lambda} = (\mathbf{I} - \mathbf{e}\mathbf{e}^T/N)(\mathbf{I} - \mathbf{e}\mathbf{e}^T/N)^T = (\mathbf{I} - \mathbf{e}\mathbf{e}^T/N).$$

Thus

$$\mathscr{E}(\boldsymbol{\zeta}^T\boldsymbol{\zeta}) = \sum_{j=1}^{N} \mathscr{E}(\zeta_j^2) = \sum_{j=1}^{N}\left(1 - \frac{1}{N}\right)\sigma^2 = (N - 1)\sigma^2.$$

On the other hand, from (10.15) we get

$$\boldsymbol{\zeta} = (\mathbf{Y} - \bar{Y}\mathbf{e}) - (\mathbf{X} - \mathbf{e}\bar{\mathbf{X}})\mathbf{b}$$

and thus

$$\mathscr{E}(\boldsymbol{\zeta}^T\boldsymbol{\zeta}) = \mathscr{E}\left\{\left[(\mathbf{Y} - \bar{Y}\mathbf{e}) - (\mathbf{X} - \mathbf{e}\bar{\mathbf{X}})\mathbf{b}\right]^T\left[(\mathbf{Y} - \bar{Y}\mathbf{e}) - (\mathbf{X} - \mathbf{e}\bar{\mathbf{X}})\mathbf{b}\right]\right\}$$

$$= \mathscr{E}\left[(\mathbf{Y} - \bar{Y}\mathbf{e})^T(\mathbf{Y} - \bar{Y}\mathbf{e})\right] - \mathbf{b}^T(\mathbf{X} - \mathbf{e}\bar{\mathbf{X}})^T(\mathbf{X} - \mathbf{e}\bar{\mathbf{X}})\mathbf{b}.$$

Equating the two expressions for $\mathscr{E}(\boldsymbol{\zeta}^T\boldsymbol{\zeta})$ and transposing the sum of squares whose value we seek, give us the first result claimed,

$$\mathscr{E}\left[(\mathbf{Y}-\bar{Y}\mathbf{e})^{T}(\mathbf{Y}-\bar{Y}\mathbf{e})\right]=(N-1)\sigma^{2}+\mathbf{b}^{T}(\mathbf{X}-\mathbf{e}\bar{\mathbf{X}})^{T}(\mathbf{X}-\mathbf{e}\bar{\mathbf{X}})\mathbf{b} \quad (10.16)$$

or

$$\mathscr{E}(\mathbf{Y}'^{T}\mathbf{Y}')=(N-1)\sigma^{2}+\mathbf{b}^{T}\mathbf{X}'^{T}\mathbf{X}'\mathbf{b}.$$

Part (b). Here we focus on the residual sum of squares, substituting the right-hand side of (10.10) for $\mathbf{Y}$, with the coefficients of the regression representation taking the values which minimize the residual sum of squares for the given data set, namely, the $\hat{a}$ and $\hat{b}$ of (10.7) and (10.9). Next we factor the matrix product and then its ensemble average, as

$$\begin{aligned}
\mathscr{E}\left[(\mathbf{Y}-\hat{\mathbf{Y}})^{T}(\mathbf{Y}-\hat{\mathbf{Y}})\right]&=\mathscr{E}\Big\{\left[(\mathbf{Y}-\bar{Y}\mathbf{e})-\mathbf{X}'(\mathbf{X}'^{T}\mathbf{X}')^{-1}\mathbf{X}'^{T}\mathbf{Y}\right]^{T}\\
&\qquad\times\left[(\mathbf{Y}-\bar{Y}\mathbf{e})-\mathbf{X}'(\mathbf{X}'^{T}\mathbf{X}')^{-1}\mathbf{X}'^{T}\mathbf{Y}\right]\Big\}\\
&=\mathscr{E}(\mathbf{Y}'^{T}\mathbf{Y}')-\mathscr{E}\left[\mathbf{Y}'^{T}\mathbf{X}'(\mathbf{X}'^{T}\mathbf{X})^{-1}\mathbf{X}'^{T}\mathbf{Y}'\right]. \quad (10.17)
\end{aligned}$$

We found the expression for $\mathscr{E}(\mathbf{Y}'^{T}\mathbf{Y}')$ in Part (a), above, so that what we need to find now is the second expectation term on the last line. Again we will use a transformation of variables and refer to Section 7.3, to express the argument $\mathbf{Y}'^{T}\mathbf{X}'(\mathbf{X}'^{T}\mathbf{X}')^{-1}\mathbf{X}'^{T}\mathbf{Y}'$ in more manageable terms and work out its ensemble average. We will need the fact that a symmetric positive-definite matrix can be written as the product of a nonsingular matrix times its transpose, which we apply to the matrix inverse term in this argument:

$$(X'^{T}X')=QQ^{T}.$$

Now we make the transformation from $\mathbf{Y}$, which has mean and covariance arrays

$$\mathscr{E}(\mathbf{Y}')=\mathbf{X}'\mathbf{b} \quad \text{and} \quad \mathscr{E}\left[(\mathbf{Y}'-\mathbf{X}'\mathbf{b})(\mathbf{Y}'-\mathbf{X}'\mathbf{b})^{T}\right]=\mathbf{I}$$

to

$$\mathbf{U}=\mathbf{Q}^{T}\mathbf{X}'^{T}\mathbf{Y}'$$

which, by the rules of transformations of Normal variables, will have mean and covariance arrays

$$\mathscr{E}(\mathbf{U})=\mathbf{Q}^{T}\mathbf{X}'^{T}(\mathbf{X}'\mathbf{b})=\mathbf{Q}^{T}(\mathbf{X}'^{T}\mathbf{X}')\mathbf{b}$$

and

$$\mathscr{C}(\mathbf{U})=(\mathbf{Q}^{T}\mathbf{X}'^{T})(\sigma^{2}\mathbf{I})(\mathbf{Q}^{T}\mathbf{X}'^{T})^{T}=\sigma^{2}(\mathbf{Q}^{T}\mathbf{X}'^{T}\mathbf{X}'\mathbf{Q}).$$

Now we note that

$$(\mathbf{X}'^{T}\mathbf{X}')^{-1}=\mathbf{Q}\mathbf{Q}^{T}\Rightarrow\mathbf{X}'^{T}\mathbf{X}'=\mathbf{Q}^{T-1}\mathbf{Q}^{-1}$$

so that

$$\mathcal{E}(\mathbf{U}) = \sigma^2 \left[\mathbf{Q}^T (\mathbf{Q}^{T-1} \mathbf{Q}^{-1}) \mathbf{Q} \right] = \sigma^2 \mathbf{I}.$$

Thus, we get both left and right equalities in the first line of

$$K\sigma^2 = \mathcal{E}\left\{ \left[\mathbf{U} - \mathbf{Q}^T (\mathbf{X}'^T \mathbf{X}')\mathbf{b} \right]^T \left[\mathbf{U} - \mathbf{Q}^T (\mathbf{X}'^T \mathbf{X}')\mathbf{b} \right] \right\}$$

$$= \mathcal{E}(\mathbf{U}^T \mathbf{U}) - \mathbf{b}^T \mathbf{X}'^T \mathbf{X}' \mathbf{Q} \mathbf{Q}^T \mathbf{X}'^T \mathbf{X}' \mathbf{b}$$

$$= \mathcal{E}(\mathbf{U}^T \mathbf{U}) - \mathbf{b}^T (\mathbf{X}'^T \mathbf{X}')\mathbf{b}$$

and consequently

$$K\sigma^2 + \mathbf{b}^T (\mathbf{X}'^T \mathbf{X}')\mathbf{b} = \mathcal{E}(\mathbf{U}^T \mathbf{U}) = \mathcal{E}\left[\mathbf{Y}'^T \mathbf{X}' (\mathbf{X}'^T \mathbf{X}')^{-1} \mathbf{X}'^T \mathbf{Y} \right]$$

which is the second term on the right-hand side of (10.17). To obtain the desired result we difference this with (10.16),

$$\mathcal{E}\left[(\mathbf{Y} - \hat{\mathbf{Y}})^T (\mathbf{Y} - \hat{\mathbf{Y}}) \right] = \left[(N-1)\sigma^2 + \mathbf{b}^T \mathbf{X}'^T \mathbf{X} \mathbf{b} \right] - \left[K\sigma^2 + \mathbf{b}^T \mathbf{X}'^T \mathbf{X}' \mathbf{b} \right]$$

$$= (N - K - 1)\sigma^2. \quad \blacksquare \qquad (10.18)$$

With the expectations of the total sum of squares and of the residual sum of squares, it is a simple step to get the expectation of the sum of squares attributable to the regression relationship. Refer to (10.11) and form the difference $\mathcal{E}(SS_{\text{total}}) - \mathcal{E}(SS_{\text{res}}) = \mathcal{E}(SS_{\text{reg}})$ to establish

COROLLARY 10.1.

$$\mathcal{E}\left[(\hat{\mathbf{Y}} - \bar{Y}\mathbf{e})^T (\hat{\mathbf{Y}} - \bar{Y}\mathbf{e}) \right] = K\sigma^2 + \mathbf{b}^T \mathbf{X}'^T \mathbf{X}' \mathbf{b}. \qquad (10.19)$$

We now have powerful equipment for testing hypotheses, as we can see when we reflect on the implications of (10.18) and (10.19) taken together. They are the expected values with respect to the distribution of the stochastic element in (10.13), of the components of the partitioned total sum of squares. We form the corresponding *mean squares* by dividing each component sum of squares by its associated degrees of freedom

$$MS_{\text{res}} = SS_{\text{res}}/DF_{\text{res}} = (\mathbf{Y} - \hat{\mathbf{Y}})^T (\mathbf{Y} - \hat{\mathbf{Y}})/(N - K - 1) \qquad (10.20)$$

and

$$MS_{\text{reg}} = SS_{\text{reg}}/DF_{\text{reg}} = (\hat{\mathbf{Y}} - \bar{Y}\mathbf{e})(\hat{\mathbf{Y}} - \bar{Y}\mathbf{e})/K \qquad (10.21)$$

and see that (10.20) is an unbiased estimate of the variance of the observation errors in the Y's and (10.21) an unbiased estimate of this same variance *plus* a positive term reflecting only the regression relationship between Y and $\mathbf{X}$.

A Test for the Significance of a Regression Relationship

A research hypothesis that proclaims the existence of a relationship between a primary variable and subsidiary variables, for which (10.13) is a valid representation for their increments, is equivalent to stating that the mean square for the regression (10.21) is estimating something larger than just the observation error variance. The additional term is

$$\mathbf{b}^T\mathbf{X}'^T\mathbf{X}'\mathbf{b}/K \qquad (10.22)$$

which will be positive unless $\mathbf{b} = \mathbf{0}$. On the other hand, the null hypothesis corresponding to this research hypothesis says that there is no such linear relationship between the increment variables, so that all the elements of the coefficient array $\mathbf{b}$ are zero. In this case the mean square in (10.21) is an estimate only of σ^2, albeit an estimate that is independent of MS_{res} in (10.20). We base our test of hypotheses on comparison of these mean squares. Specifically, we will base it on the ratio

$$MS_{reg}/MS_{res} = \left[(\hat{\mathbf{Y}} - \bar{Y}\mathbf{e})^T(\hat{\mathbf{Y}} - \bar{Y}\mathbf{e})/K\right]\bigg/\left[(\mathbf{Y} - \hat{\mathbf{Y}})^T(\mathbf{Y} - \hat{\mathbf{Y}})/(N - K - 1)\right].$$

If the research hypothesis is true, the numerator of the ratio is estimating something larger than what the denominator is estimating. If there is no linear regression relationship between Y and $\mathbf{X}$, then $\mathbf{b}$ will be a vector of zeros, and the numerator and denominator will simply be two estimates of the same observation error variance σ^2. In the latter case we would expect the ratio to be close to 1.0. Evidence of the truth of the research hypothesis will present itself as a value for the ratio in excess of 1.0: *The larger the ratio, the greater the credibility of the hypothesized regression relationship.*

The remaining link in constructing a test for the significance of an observed ratio of mean square errors is the distribution of the ratio on the assumption that $\mathbf{b} = \mathbf{0}$. This distribution is obtained as an immediate consequence of the following two corollaries.

COROLLARY 10.2.   *When* $\mathbf{b} = \mathbf{0}$,

$$SS_{total}/\sigma^2 = (\mathbf{Y} - \bar{Y}\mathbf{e})^T(\mathbf{Y} - \bar{Y}\mathbf{e})/\sigma^2$$

has Chi Square distribution with $(N - 1)$ *degrees of freedom.*

Proof.   Here we use the representation (10.13) for $\mathbf{Y}$ with the attendant assumption that

$$\boldsymbol{\varepsilon} \to \mathcal{N}(\mathbf{0}, \sigma^2\mathbf{I}).$$

Applying (8.4) to the Normal variable $\boldsymbol{\varepsilon}$, we have

$$\boldsymbol{\varepsilon}^T\boldsymbol{\varepsilon}/\sigma^2 \to \chi_N^2.$$

And now, partitioning $\boldsymbol{\varepsilon}^T\boldsymbol{\varepsilon}$ as

$$\sum_{j=1}^{N} \varepsilon_j^2 = \sum_{j=1}^{N} \left(\varepsilon_j - \bar{\varepsilon}\right)^2 + N\bar{\varepsilon}^2$$

we can use the result given by Assertion 8.1, again applied to $\boldsymbol{\varepsilon}$, to get

$$\sum_{j} \left(\varepsilon_j - \bar{\varepsilon}\right)^2/\sigma^2 \to \chi^2_{(N-1)} \quad \text{and} \quad N\bar{\varepsilon}^2/\sigma^2 \to \chi^2_{(1)},$$

where these two terms are stochastically independent. In the present context, another way of writing the first equivalence is

$$\boldsymbol{\zeta}^T\boldsymbol{\zeta}/\sigma^2 = (\boldsymbol{\varepsilon} - \bar{\varepsilon}\mathbf{e})^T(\boldsymbol{\varepsilon} - \bar{\varepsilon}\mathbf{e})/\sigma^2 \to \chi^2_{(N-1)},$$

where we are using $\boldsymbol{\zeta}$ as it is defined in Part (a) of our proof above. Now we need only take $\mathbf{b} = \mathbf{0}$, to have

$$\boldsymbol{\zeta} = (\mathbf{Y} - \bar{Y}\mathbf{e}) - (\mathbf{X} - \mathbf{e}\bar{\mathbf{X}})\mathbf{b}\big|_{\mathbf{b}=\mathbf{0}} = (\mathbf{Y} - \bar{Y}\mathbf{e})$$

and thus

$$(\mathbf{Y} - \bar{Y}\mathbf{e})^T(\mathbf{Y} - \bar{Y}\mathbf{e})/\sigma^2 \to \chi^2_{(N-1)}. \qquad \blacksquare$$

CorOLLARY 10.3.   *When* $\mathbf{b} = \mathbf{0}$,

$$SS_{reg}/\sigma^2 \quad \text{and} \quad SS_{res}/\sigma^2$$

are independent Chi Square variables, the first with K degrees of freedom and the second with $N - K - 1$ degrees of freedom.

Proof.   We use Corollary 10.2 and recall Cochran's Theorem (8.2), since

$$SS_{total} = SS_{reg} + SS_{res} \quad \text{and} \quad DF_{total} = DF_{reg} + DF_{res}. \qquad \blacksquare$$

Now we invoke (8.18), which gives us the distribution of the ratio of standardized variables, and immediately get the result that is central to testing regression hypotheses:

LEMMA 10.1.   *When* $\mathbf{b} = \mathbf{0}$,

$$\left[SS_{reg}/K\right]\big/\left[SS_{res}/(N-K-1)\right]$$

has Fisher's F distribution with numerator and denominator degrees of freedom K and $(N - K - 1)$, respectively.

10.5 TESTING REGRESSION HYPOTHESES

Now we are positioned to assess the *significance* of the linear regression for observed relative humidity, with daily observations of RH, pressure, average temperature, temperature change, and wind speed, in so far as a

TABLE 10.4 Analysis of Variance for RH with C2, C4, C7

SOURCE	DF	SS	MS	F	P
REGRESSION	3	2069.70	689.90	7.91	0.001
ERROR	26	2267.67	87.22		
TOTAL	29	4337.37			

with regression equation

$$\widehat{RH} = 1428 - 44.8 \text{ PRESSURE} - 0.399 \text{ DELT} - 1.94 \text{ WND SPD}.$$

linear expression in subsidiary variables can account for the observed variability of RH. (Review Table 10.3 which summarizes the results of the four analyses, with successive additions of subsidiary variables to the regression. Results of the regression for RH on the increments of the three variables, PRESSURE, DELT, and WND SPD, are shown in Table 10.4. This analysis of variance table gives us the final result of the test for significance of the F ratio

$$F = MS_{reg}/MS_{res}$$

for the hypotheses

$$H_0: \mathbf{b} = \mathbf{0} \qquad \text{versus} \qquad H_R: \mathbf{b} \neq \mathbf{0},$$

together with the bookkeeping details of the analysis. It reports total and component sums of squares, their associated degrees of freedom, and the mean squares from which the F ratio is formed, along with F and its P value.

By Lemma 10.1, the F ratio will have Fisher's F distribution if $\mathbf{b} = \mathbf{0}$. The corresponding *P-value*, in the ANOVA table, is *the probability of observing a value as large as the one produced by the analysis, with the associated DFs, when the distribution of the test statistic is Fisher's F*. Thus a very small value for P indicts the hypothesis "$\mathbf{b} = \mathbf{0}$".

Now consider F in the context of the research hypothesis and recall that the numerator of the F ratio estimates the same variance value as the denominator estimates *plus* the term given by (10.22):

$$\mathbf{b}^T \mathbf{X'}^T \mathbf{X'} \mathbf{b}/K.$$

Were H_0 true, this term would be 0. However, when $H_R: \mathbf{b} \neq \mathbf{0}$ is true, this is a positive addition to the variance estimate, because $\mathbf{b}^T \mathbf{X'}^T \mathbf{X'} \mathbf{b}$ is a sum of squares. Then the numerator of F is estimating something larger than the denominator is estimating; and the ratio does not have Fisher's F distribution. Its distribution is stochastically larger. Accordingly, large values of F, with small values of P, support the conclusion that the research hypothesis is true, namely that a linear combination of increments of the subsidiary variables explains a significant portion of the variability of the primary variable from its own mean.

The *P*-value is essentially the bottom line of the analysis. Another way of phrasing its definition is as *the probability that the evidence of a linear relationship is only an artifact of chance association*. If we have preselected a size for the test of significance, say α, then we compare the output value of *P* with α. If *P* is less than α, then of course we reject H_0 and accept H_R. If *P* is greater, then we cannot reject H_0 at level α. When the *F* ratio MS_{reg}/MS_{res} is not large enough to support the conclusion that the MS_{reg} is estimating something larger than MS_{res}, that is, that $\mathbf{b} \neq \mathbf{0}$, we will have to choose either to abandon our objective or accumulate additional evidence if we are committed to the research hypothesis.

With the results shown in Table 10.4, $P = 0.001$; and no reasonable test of hypotheses would fail to reject H_0. We conclude with confidence that a linear expression for relative humidity in terms of daily pressure, temperature change, and wind speed, does account for a significant fraction of the observed variability of RH values ($\sim 48\%$, by Table 10.3).

We will consider the analysis of variance table for the regression which adds average temperature to the previous three subsidiary variables, to illustrate the effect of including an additional variable. This is Table 10.5, and we can see that the values of the statistics presented there are very close to those of Table 10.4. The corresponding coefficients of the two regression equations are similar as well, with the additional variable receiving only small weight. In Table 10.5 the degrees of freedom for the regression sum of squares is larger by one and the *DF* for the residual sum of squares is smaller, because of the inclusion of one additional variable in the analysis. Although this inclusion did not add significantly to the regression fit, its coefficient was estimated from the data and a degree of freedom was lost from the error variance estimate. In fact, the "explained sum of squares" is smaller here than in Table 10.4, as is the value of *F*. Thus we have detracted a little from the precision of the representation for RH, by the inclusion of a variable which has a very weak linear

TABLE 10.5 Analysis of Variance for RH with C2, C3, C4, C7

SOURCE	DF	SS	MS	F	P
REGRESSION	4	2071.11	517.78	5.71	0.002
ERROR	25	2266.26	90.65		
TOTAL	29	4337.37			

with regression equation

$\widehat{RH} = 1433 - 44.9 \text{ PRESSURE} - 0.024 \text{ AVETEMP} - 0.399 \text{ DELT} - 1.94 \text{ WND SPD}.$

relationship with our primary variable: The regression relationship is more effective without it.

Some Final Notes

Beyond what we have presented and discussed so far, you will have discovered that the output for MINITAB regression analyses produces a table of statistics for the subsidiary variables individually. As an example, the contents of Table 10.6 came with the output of the analysis we have just considered. The table contains the regression coefficients for each of the variables and with the coefficients are associated values in columns labeled STDEV, T-RATIO, and P. The latter may be used as *indicators* of the stabilities of the regression coefficients, with reference to a possibly enlarged data set or a different set of observations. In fact, these statistics may identify variables that might well be omitted from the regression, such as average temperature in the present regression. *What the information in this table does not do* is to give us reliable significance levels for individual coefficients. It is true that a large value in the P column indicates that the corresponding subsidiary variable does not make a reliable contribution to the regression. However, we cannot interpret these *P*-values as independent, individual, levels of significances of the predictor coefficients; because they have been estimated simultaneously from a common set of observations of variables whose interrelationships (or, interdependencies) are the core of the analysis. We recommend that they be used as indicators only, without claiming the suggested precisions.

Simultaneous confidence intervals for the regression coefficients and tests for their individual differences from zero, with collective measures of confidence and significance, would be valuable additions to the analysis. These are difficult, but technically possible to obtain. See Scheffé (1959, pp. 68–83) for a presentation of the theoretical structure for simultaneous confidence intervals. Generally the work required to construct valid simultaneous intervals outweighs the perceived need for them.

TABLE 10.6   Details of the Analysis of Variance for RH with Four Predictors

PREDICTOR	COEF	STDEV	T-RATIO	P
CONSTANT	1433.	338.	4.24	0.000
PRESSURE	−44.9	11.2	−4.03	0.000
AVETEMP	−0.024	0.192	−0.12	0.902
DELT	−0.401	0.156	−2.57	0.016
WNDSPD	−1.97	0.760	−2.60	0.015

EXERCISES

1. (a) For each of the three examples in the second paragraph of this chapter, identify primary and subsidiary variables, as you would use them in a regression analysis.
   (b) For one of these examples, describe a hypothetical data set such as you could use to estimate coefficients in a linear regression representation.
   (c) For part (b), identify the elements of the arrays of observed values **Y** and **X**, as these would be used in the regression analysis.

2. Focus on the third example, and take PV = 700 mb wind speed.
   (a) Identify a plausible grid for a Gulf of California analysis.
   (b) Describe the sources of observations, in terms of variables, instrumentation, and plausible locations, which could go into such an analysis.
   (c) Identify a "guess field" that could be used to define increments for the wind and geo-potential variables for this linear regression analysis, at the locations for which you have observed values for these variables.
   (d) For part (c), what would you anticipate the ensemble means of the increments to be? How could you confirm that this is a reasonable expectation?

3. Use the subfile MAY.1989 created for Crookston, MN, in Chapter 2.
   (a) Choose CRKSMIN to be the PV and CRKSMAX to be the SV. Identify and find the values of each of the terms $\bar{Y}$, $\bar{X}$, S_{YX}, and S_X^2, in the final equation of (10.3).
   (b) Use a statistic based on the correlation, from Chapter 8, to establish whether or not you believe there is a significant linear relationship between CRKSMIN and CRKSMAX.
   (c) Present a sound supporting argument for your conclusion in part (b).
      Hint: Why does $\rho = 0$ imply $b = 0$?

4. Use the subfile MAY.1989 of Exercise 3.
   (a) Make two new columns containing the differences between the daily values and the May 1989 means for CRKSMIN and CRKS-MAX. Name these new columns INCPV and INCSV, and print all four columns. Label the printed table, and cut it out and stick it on your homework sheet.
   (b) Plot INCPV versus INCSV. Label this plot clearly enough that you will recognize it a month from now, and place it on your homework sheet. Describe the array in words, as you see it. (There's no right/wrong description here.)

(c) Make two photocopies of the plot of part (b). On one, draw and label the line $Y = \bar{Y}$; and on the other draw a line roughly where you think the regression line is likely to go. Use these plots to dot-in the residuals, as done in Figs. 10.1(a) and 10.1(b).

(d) Calculate the total and residual sums of squares, and then the sum of squares due to regression. Report and label these clearly. Explain how you got them in sufficient detail that I could reproduce your calculations without bothering you with a phone call.

5. Use the subfile MON.CHGS created from STN#27 in Chapter 2, Exercise 4(b).

   (a) Select 50-m monthly change data for temperature and salinity, and place it in two columns named DELT versus DELS. Plot DELT versus DELS, and place a clearly labeled copy in your homework sheet.

   (b) Use a MINITAB regression command to find the regression coefficients for

   $$\widehat{DELT} = \hat{a} + \hat{b}\, DELS.$$

   Report both the command you used and the resulting values of $\hat{a}$ and $\hat{b}$.

   (c) With ruler and pencil, plot the regression equation over the plot of part (a).

   (d) Are there any large residuals identified by the analysis? If so, identify them with the variable and month of occurrence, and circle the corresponding point(s) on the plotted array.

   (e) In the MINITAB output find the values for $SS_{total}, SS_{res}, SS_{reg}$, and report these on your homework paper.

6. Create a new MINITAB subfile of five columns, with data from STN#27. Put month numbers in C1, 10-m temperatures in C2, 10-m salinities in C3, surface temperatures in C4, and surface salinities in C5. Name the columns MONTH, 10MTEMP, 10MSALT, 0MTEMP, 0MSALT. Call this file TRAWLN and save it.

   (a) Print the file TRAWLN and put it in your homework.

   (b) Write the MINITAB command you would use for a regression analysis with 10-m temperature as the PV and *two* subsidiary variables: surface temperature and surface salinity.

   (c) In the matrix expression $\mathbf{Y}' = \mathbf{X}'\mathbf{b}$, describe the elements of each of the terms in words. Then give the orders of the arrays and identify these with the factual characteristics of the data set (numbers of observations and numbers of variables).

7. Apply the MINITAB command of Exercise 6(b) to the subfile TRAWLN and annotate the output, to the extent that you are able to identify each of its components without reading ahead or elsewhere.

8. Present a rigorous proof that (10.9)

$$\mathbf{b} = (\mathbf{X}'^T \mathbf{X}')^{-1} \mathbf{X}'^T \mathbf{Y}'$$

does indeed minimize the total sum of squares of (10.8)

$$SS = (\mathbf{Y}' - \mathbf{X}'\mathbf{b})^T (\mathbf{Y}' - \mathbf{X}'\mathbf{b}).$$

9. Change the MINITAB command of Exercise 6(b) to include a third subsidiary variable, namely 10MSALT.
   (a) Copy and compare the ANOVA tables for this and the two-SV analysis of Exercise 7.
   (b) For the analysis with three SVs, why are the degrees of freedom for the residual sum of squares equal to 8?
   (c) What are comparable estimates for observation error variance, from the two analyses? Which has the larger degrees of freedom and why?
   (d) Write the mathematical expression for the fraction of the total variation of 10-m temperatures "explained by regression" and the designation for this value in MINITAB output. Compare the fractions of "explained variability" for the two regressions, with comments.

10. Select 90 days data, for the fall of 1989, from the file CLIMAT.
    (a) Choose peak wind speed as the primary variable and identify the three variables most strongly correlated with it. Show your evidence.
    (b) Run a MINITAB regression for peak wind speed, with the three variables of part (a) as subsidiary variables. Place the ANOVA table in your homework paper.
    (c) What must you assume about the distribution of $Y =$ peak wind speed, in order to use the results of statistical theory to test the hypotheses
        H_R: Variability of the PV is partially accounted for by a linear relationship with the CVs; versus
        H_0: There is no linear relationship that "explains" variability of peak wind speed in terms of other variables.
    (d) If there is a linear regression relationship, what is the only term that gives you an unbiased estimate of the variance (σ^2) of the observation error for peak wind speed?

11. (a) Refer to the text example using the data in Table 10.1. If the true value of the regression coefficient b is 0.55, how much bigger than the observation error variance is the value estimated by the mean square in the numerator of the F statistic?
        Hint: You must calculate this from the data, using the ensemble average of the mean square as a guide.

(b) What do you estimate the comparable value to be for the peak wind speed analysis of Exercise 10? Give the formula for your estimate. Do you think you can distinguish the influences of the subsidiary variables?

12. Follow the text example of a linear regression analysis with several variables, using dew point temperature (DPT) in place of relative humidity (RH). Provide a clear description of what you have done, especially the choices you make in the course of analysis. State H_R and H_0. Present the evidence relevant to testing the hypotheses, your conclusion, and supporting arguments.

11

BOOTSTRAPPING
"scientific inference
when none
of the above apply"

11.1 INTRODUCTION

The preceding chapters have introduced some high-powered tools of scientific inquiry. However, these can only be useful when their application is valid. Although following the development of procedures for statistical estimation and testing as closely as we have may have seemed tedious (occasionally in the extreme), it is essential that we know when we are on solid ground. The validity of an analysis is the keystone of clear presentation and strong defense of results and conclusions. If we can establish it, we're home. If we cannot, then we must base scientific inferences on alternative procedures. This final chapter is concerned with management of uncertainty and inference in research settings in which the use of classical statistical analysis tools would be invalid.

Throughout the text we have dealt with the analysis of scientific data for systems for which we can generally articulate physical and chemical relationships via mathematical models and computer algorithms. We have treated the discrepancies between observed values of variables which are used to describe states of these systems and values that would satisfy known relationships, as *stochastic variables*. And we have used the formal structure of Statistics to make inferences in the resolution of scientific issues concerning these discrepancies.

Thus far we have reviewed *classical distributions* of traditional statistics and explored their potential for making inferences about features of observed systems. In the course of the foregoing review and exploration, we have emphasized assumptions that give validity to the inference mechanisms we have studied. In many places, the Normal distribution has played a major role, either as a univariate distribution or as a multivariate distribution of several increment variables considered in relation to one another. In Chapter 7, we referred to and loosely paraphrased *the Central Limit Theorem*: a theorem which explains why so much emphasis is placed on the Normal distribution and on distributions for test statistics derived from the Normal. As we have seen, the χ^2, Student's t, and Fisher's F distributions enable construction of tests of hypotheses and of confidence intervals, for means and variances, and for differences between means and ratios of variances, *given that the underlying assumptions are valid*. However, if the assumptions are invalid, the inferences constructed from these distributions are invalid. Then we must use other mechanisms to obtain the required answers.

In this chapter, we are concerned with situations in which the basic assumptions of the Normal distributions may not apply. However, of even greater concern are circumstances in which observed increment variables do not satisfy the additional assumption of sequential independence which would justify straightforward use of the χ^2, t, or F distributions.

Unifying characteristics of all the fields studied in atmospheric and ocean sciences are their spatial and time coherence. In fact they would not be interesting and their scientific study would not be possible, if they were not *spatially and temporally coherent*. Nonetheless *it is this feature that frequently thwarts justification of the use of classical statistical inference techniques* because it can so easily invalidate the assumption of the stochastic independence of the elements of a data set. Observed increments may have the coherence of either the "guess field" that defines their background or the observation error from common recording and transmission equipment, or both.

If the assumption of stochastic independence is invalid, then how do we construct confidence intervals and tests?

11.2 BOOTSTRAP TECHNOLOGY

There is a growing class of procedures that underwrite statistical estimation, diagnostic analyses, and testing hypotheses with data for which the individual values do not have the stochastic independence required by classic statistical estimation, diagnostics, and testing. The procedures may be classified as either *resampling* or *stochastic simulation*. If you have done a bit of reading in this area you may be tempted to associate the concept of bootstrapping with resampling, but not with stochastic simulation. However, the techniques in both of these divisions proceed from the same principle and may be thought of collectively as bootstrapping. The unifying idea is that we can use available data to describe the unknown distribution of the chosen test statistic, or characteristics of it, which can then serve as an inferential base. On this base, computer-simulated, stochastic sampling is applied to generate confidence intervals or tests of hypotheses. Thus the data analysis is *lifted* (*from the data*) *by its own bootstraps*.

We use the term *bootstrap technology* to characterize the generic class of estimation and testing procedures that base their construction on information in the data set itself, without involving external distribution assumptions. The core of bootstrapping is approximation of the probability mechanism that generated the data, followed by the use of the approximation in simulating a large number of pseudo observations. These methods require fast, efficient computation capability. However, high-speed computing facilities are now part of our basic scientific equipment and the costs for their use are standard line items in any science budget. Only lack of familiarity and shortage of good, user-friendly software inhibit full use of these powerful, inferential tools. As computational support for bootstrapping technology is further developed, these tools will become increasingly central to the formulation of inferences in response to scientific questions that must be answered with the interdependent data of dynamic systems.

Of the two subdivisions of bootstrapping procedures, *resampling* means *re-using the elements of the initial set of observations in a hypothetical experimental setting in which we "observe" many randomly selected subsets*. Estimates of system parameters, probabilities, and confidence intervals for predictions, are all based on the outcomes of the many, pseudo observation sets. To describe this we shall denote the unknown probability mechanism governing the original observations by F, and the estimate for this mechanism obtained as the observed or "empirical" distribution of the data by $\hat{F}$. The "resampling" which is the random selection from among the original observations, uses the distribution defined by $\hat{F}$ as the template for random selections. (See Efron and Tibshirani (1986) for a good introduction to this bootstrap technique.) From the viewpoint of statistical theory, resampling provides a sound basis for inference in the sense of unbiased conditional inferences. From the viewpoint of a skeptic, it requires the faith that the original set of observations did not describe a maverick situation whose characteristics will distort its subsamples. From the viewpoint of a pragmatist, it is the best we can do with the information we have.

Stochastic field simulation (or SFS) is distinct from resampling. As SFS is applied in the analysis of stochastic features of geophysical variables, the increment field is assumed to have a low order, spatial/temporal autoregressive structure. The parameters of this structure, such as the spatial or temporal autocovariance and the variance of white noise innovations, are estimated from the original set of observations. (See Thiébaux and Pedder, 1987, and Cressie, 1991, for descriptions of some techniques of estimation and parametric modeling.) Then a null hypothesis version of the observed field is reproduced a very large number of times. This may be done either by generating a deterministic field and, for each simulation, adding a new set of random errors at the locations corresponding to observations, or by literally reproducing a multidimensional autoregressive process, with each reproduction driven by different, randomly generated white noise innovations. (See Franke, 1985; Zwiers, 1987; and Zwiers and Thiébaux, 1987, for brief discussions of such field generation, and Ripley, 1987, for a more general treatment of stochastic simulation.) Each generated field is "observed" at locations corresponding to those of the true data set, and for each set of pseudo observations, a statistic or array of statistics is computed. Whatever the focus of the analysis, e.g., a suspected regional bias or climatological trend, the statistic or array of statistics chosen to measure it is computed from the simulated, null hypotheses fields, thus generating its distribution for the test of hypotheses.

Technical reviews and reports on continuing development of theory and computational techniques for bootstrapping are given by DiCiccio and Romano (1988), Hinkley (1988), Hall and Pittelkow (1990), and Leger, Politis, and Romano (1992), among others. The following section presents a few examples of applications in atmospheric science.

11.3 EXAMPLES OF BOOTSTRAPPING IN GEOPHYSICS

Resampling Applied to Comparison of Polar and Tropical Satellite Radiances

Crone and Crosby (1993) have applied to bootstrap technology to study regional differences in upwelling radiances measured by the polar orbiting meteorological satellite TIROS-N. The authors used a data base of over 360 measurements from the polar regions (60°–90°) and from the tropical region (30°S–30°N), from which they generated distributions of the square of a *distance metric* for statistical comparison of observed eigenvectors of covariance matrices, for the contrasting regions. The metric is the *distance between the two 4-dimensional subspaces spanned by the principal components computed from radiance vector data* from polar and tropical regions. The distributions of the metric were computed in two ways: for the

TABLE 11.1 Bootstrap Estimates of Means and Standard Deviations of the Squared Distance between 4-Dimensional Subspaces Spanned by Principal Components

	$\hat{\mu}_{D^2}$	$\hat{\sigma}_{D^2}$
Polar /tropical	1.007	0.026
Tropical/tropical	0.049	0.033

polar/tropical contrast, by resampling from the two distinct, original sets of observed vectors, and for the "no difference" hypothesis, by drawing both pseudo subsamples from the tropical data set. In each case, the resampling was done 500 times. Means and standard deviations of the generated distributions were computed for comparison. Table 11.1 presents some of their results, which support the conclusion that upwelling radiances are significantly different in polar and tropical regions.

Test for Correlation between Climate and the 11-Year Solar Cycle

Barnston and Livezey (1991) used a stochastic simulation technique to generate a "no correlation" background distribution of a statistic applied in their study of the evidence for correlation of the solar flux and the Northern Hemisphere 700-mb height field, for a 40-year period 1951–1990. In constructing a test for correlation between solar flux and 700-mb height, the authors have used a familiar summary statistic adjusted for the time correlation of the data,[1] as the basis for the hemispheric, correlation impact statistic.[2] Bootstrap generation of the distribution of the statistic, in place of the classical statistical distribution, was required for evaluation of its significance because of the cyclical character of the solar flux and the spatial dependence of the climate fields. The former creates interannual autocorrelation in solar flux observations and the latter creates large spatial correlations among neighboring gridpoint data for the height field. Together these temporal and spatial correlations in the data nullify assumptions basic to the classical distribution of the test statistic. Consequently its distribution was simulated by performing 2000 random shufflings of the phase assignments to the data with recalculation of the test statistic for each shuffle. The procedure preserved the influence on the test statistic of the temporal and spatial correlations in the data base, while negating the research hypothesis of correlation between flux and

---

[1] $t = [r^2(N-2)/(1-r^2)]^{1/2}$, where N is a calculated "number of independent time points".
[2] See Davis (1976) and Thiébaux and Zwiers (1984) for a description of the use of this statistic and discussion of questions relating to its use.

700-mb heights. Thus the generated distribution was a true null hypothesis distribution for the test statistic.

Statistical Analysis of Model Climate Simulation

Livezey (1985) describes the use of bootstrap techniques (permutation or Monte Carlo simulation) in evaluation of general circulation model experiments. Because the data sets are small collections of large numbers of cross-correlated climate statistics, they fail the criteria of classical statistical test procedures. For example, the author refers to a prescribed change experiment for which the 30-day mean circulation is compared to that from a 6-case mean control, i.e., 6 runs of the same model without the intervention of the "prescribed change". To evaluate the difference in outcome, 9 additional (unmatched) model runs without the prescribed change are identified, to make a total of 15 cases from which 30-day means can be computed. The null hypothesis distribution of the difference between the original "prescribed change run" and the 6-case mean control is generated by repeatedly, randomly selecting 7 at a time from the 15 and comparing 6 of these to the 7th. Since there are more that 40,000 different possible comparisons, the null hypothesis distribution corresponding to "no change effect" could be constructed from a large number of distinct random selections.

Generation of Confidence Intervals for the Correlation between the Southern Oscillation Index and January-to-March Rainfall over Central South Africa

Mason and Mimmack (1992) obtain confidence intervals for the correlation of the Southern Oscillation Index (SOI) and Southern Hemisphere summer rainfall, with data from 59 stations in South Africa. In their paper the authors discuss the importance of, and the mechanisms for, generating confidence intervals for characteristics of observed climatological systems, where these characteristics are expressed as parameters of stochastic distributions. And they adapt this discussion to the particular problem of a correlation estimate: which is neither unbiased nor normally distributed.

In the case of assumed non-zero correlation, the distribution of the correlation estimate is not known. Mason and Mimmack describe the generation of bootstrap confidence intervals for correlation of the two climatological variables, using the observed correlation coefficient calculated from a long data record, together with a distribution obtained by resampling from the data record. With this distribution, the authors estimated both the bias in the original correlation estimate and interval confidence bounds. They did this for the full 48-year data set 1935–1983; and repeated the procedure for data of the more recent 30-year period

TABLE 11.2 Ninety-Percent Bootstrap Confidence Intervals for r^2, the Coefficient of Determination for the Southern Oscillation Index and January-to-March Rainfall in Central South Africa

Data set	Lower limit	Upper limit
1935–1983	11%	36%
1954–1983	16%	54%

1954–1983, known to have greater stability in the SOI/rainfall relationship. Table 11.2 illustrates their results in terms of *the coefficient of determination*, r^2: the percentage of the observed variance of one variable which may be associated with or attributed to an associated second variable.

11.4 OPEN PROBLEMS IN THE DEVELOPMENT AND USE OF BOOTSTRAP ANALYSIS TECHNIQUES

Ramage (1983) and Brown and Katz (1991) raise problems connected with the use of the correlation coefficient as a statistical tool in climatology, which are not addressed by the foregoing. Solow (1985) and Zwiers (1990) are specifically concerned with the use of resampling procedures in any geophysical context where the data set variables are auto-serially correlated. Solow points out that the bootstrap as formulated by Efron (1982) cannot be applied to serially correlated data. Zwiers discusses the effect of serial correlation on statistical inferences made with resampling procedures that do not specifically accommodate auto-serial correlation of the variables observed; and he also illustrates the effect of serial correlation on inferences made with bootstrap resampling in the form in which it was originally proposed. To do this, Zwiers simulates stochastic autoregressive processes to create the data set which he uses for illustration. In his first example, he carries out a large-scale simulation experiment to determine the true significance levels for difference-of-means tests using a bootstrap permutation technique in which he assigns (known) lag-correlation values to the simulated processes. His results for nominal significance level $\alpha = 0.05$, shown in Fig. 11.1, make the point that an assumed level of significance for a test of hypotheses will be underestimated, with serious underestimates corresponding to strongly correlated data.

In a second example, Zwiers illustrates effects of serial correlation in bootstrap confidence intervals constructions. He has calculated large sample estimates of the true probabilities that nominal 95% bootstrap confidence intervals include the values of the parameters for which they are

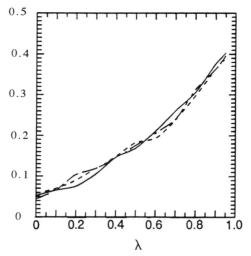

FIGURE 11.1   True, calculated significance of a permutation difference-of-mass test with nominal significance $\alpha = 0.05$, plotted versus the lag-correlation of the simulated stochastic series used in the means test.

constructed.[3] His results for confidence intervals for differences-of-means are shown in Fig. 11.2, where the three curves correspond to $n = 20$, 40, and 100 sequential "observed" values of the simulated time series. Here again it is clear that serially correlated data can lead to a serious underestimate of the probability that a bootstrap confidence interval contains the true mean difference.

These examples, and other illustrations of impact on measures of significance and confidence in estimates brought through the natural coherence of geophysical systems, have addressed the *temporal* coherencies of time series. The problems presented to inference constructions by serial correlation of time series will have *parallels in larger dimensions for the analysis of spatially correlated data arrays*. We do not have ready solutions for these problems. They are raised here, at the conclusion of this text, because they cannot be neglected in drawing conclusions from statistical analyses. Interpretation of results of analyses made with auto-correlated data must take into account the impact of the auto-correlation on the test statistics. Solow has described a technique to accommodate serial correlation in the analysis of some time series. Continuing research

---

[3]This was done by replicating the process of obtaining a pair of "observed time series" and constructing the bootstrap confidence intervals 1000 times, using simulated stochastic autoregression series to generate "data".

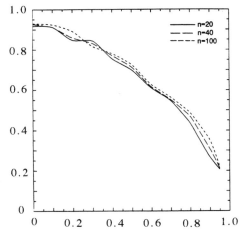

FIGURE 11.2   The true probabilities of "coverages" of nominal 95% bootstrap confidence intervals, for differences between means of two observed series, for first-order autoregressive processes, versus the serial-lag-correlation λ. The solid line records the coverage for $n = 20$, the long dashed line for $n = 40$, and the short dashed line for $n = 100$ sequential values in each simulated series.

will deal with its extension to spatial dimensions. The problems are tough and resistant; their solutions are critical to sound scientific inference.

EXERCISES

1. "Scientific Question Revisited". Refer to Exercise 1.2, in which you were asked to formulate a research question, providing the following information.
   (a) A statement of the scientific objective;
   (b) Your knowledge or assumptions about sources of uncertainty and measurement errors;
   (c) A description of how the data was, or might be obtained;
   (d) Your expectation of the use of the data in answering the research question(s) of the scientific objective.

   Retrieve your copy of this exercise and critique it, in terms of what you now know about making inferences when you have observations of an incompletely known system and the observations contain uncertainties.

   What you are being asked to do here is to go back to what you were thinking when you formulated the original research objective; treat it as if it were someone else's formulation; and help them reformulate it in terms of inferences and hypotheses that can be handled with the statistical analysis techniques that you have learned. Make the data specifications precise and state the assumptions you will make about the variables you have (perhaps hypothetically) observed.

This exercise will provide a mechanism for review of data management requirements, and the capabilities and constraints of statistical inference as a framework for scientific thinking.

2. Find two, contrasting papers in the scientific literature and critique their analyses of data. Use what you have learned to argue for their validity or invalidity. If your assessment of either of them is that the analysis or conclusions are weak, append a detailed outline for a recommended alternate analysis.

REFERENCES

Barnston, A. G., and R. E. Livezey (1991): The statistical testing of associations between the 10.7 cm solar flux, the QBQ and the lower tropospheric climate in the Northern Hemisphere in winter. *J. Geomag. Geoelectr.*, **43** (Supplement), 731–740.

Brown, B. G., and R. W. Katz (1991): Use of statistical methods in the search for teleconnections: Past, present, and future. *In* "Teleconnections Linking Worldwide Climate Anomalies: Scientific Basis and Societal Impact", pp. 371–400, Cambridge Univ. Press, Cambridge.

Brownlee, K. A. (1965): "Statistical Theory and Methodology in Science and Engineering", Wiley, New York, 590pp.

Conover, W. J. (1980): Statistics of the Kolnogorov–Smirnov type. *In* "Practical Nonparametric Statistics", 2nd ed., Chap. 6, Wiley, New York, 493pp.

Cressie, N. A. C. (1991): "Statistics for Spatial Data", Wiley, New York, 900pp.

Crone, L. J., and D. S. Crosby (1993): Statistical applications of a metric on subspaces to satellite meteorology. *In* "Proceedings, Interface 93: Computing Science and Statistics, 14–17 April 1993, Dan Diego, CA".

Davis, R. (1976): Predictability of sea surface temperature and sea level pressure anomalies over the North Pacific Ocean. *J. Phys. Oceanography*, **6**, 249–266.

D'Agostino, R. B., A. Belanger, and R. B. D'Agostino, Jr. (1990): A suggestion for using powerful and informative tests of normality. *Amer. Statist.*, **44**, 316–321.

Diaconis, P., and B. Efron (1983): Computer-intensive methods in statistics. *Sci. Amer.*, **48**, 116–130.

DiCiccio, T., and J. Romano (1988): A review of bootstrap confidence intervals (with discussion). *J. Roy. Statist. Soc. Ser. B*, **50**, 338–370.

Draper, N. R., and H. Smith (1981): "Applied Regression Analysis, 2nd Edition", Wiley, New York, 709pp.

Efron, B. (1982): "The Jackknife, the Bootstrap and Other Resampling Plans", J. Arrowsmith, 92pp.

Efron, B., and R. J. Tibshirani (1986): Bootstrap methods for standard errors, confidence intervals and other measures of statistical accuracy. *Statist. Sci.*, **1**, 54–77.

Fisher, Sir R. A., and F. Yates (1967): "Statistical Tables for Biological, Agricultural and Medical Research", Hafner, New York, 146pp.

Franke, R. (1985): Sources of error in objective analysis. *Monthly Weather Rev.*, **113**, 260–270.

Glen, S. M., D. L. Porter, and A. R. Robinson (1991): A synthetic geoid validation of geosat mesoscale dynamic topography in the Gulf Stream region. *J. Geophys. Res.*, **96**, 7145–7166.

Graybill, F. A. (1961): "An Introduction to Linear Statistical Models", McGraw–Hill, New York, 463pp.

Graybill, F. A. (1983): "Matrices with Applications in Statistics", Wadsworth International Group, Belmont, CA.

Hall, P., and Y. E. Pittelkow (1990): Simultaneous bootstrap confidence bands in regression. *J. Statist. Comput. Simulation*, **37**, 99–113.

Hinkley, D. (1988): Bootstrap methods. *J. Roy. Statist. Soc. Ser. B*, **50**, 321–337.

Kendall, M. G., and A. Stuart (1963): "The Advanced Theory of Statistics, Vol. 1: Distribution Theory", Griffin, London, 433pp.

Kendall, M. G., and A. Stuart (1967): "The Advanced Theory of Statistics, Vol. 2: Inference and Relationship", Hafner, New York, 690pp.

Kotz, S., and N. L. Johnson (Eds.) (1982: "Encyclopedia of Statistical Sciences, Vol. 1: A to Circular Probable Error", Wiley, New York, 480pp.

Lange, M. A., and H. Eicken (1991): The sea ice thickness distribution in the northwestern Weddel Sea. *J. Geophys. Res.*, **96**, 4821–4837.

Leger, C., D. N. Politis, and J. P. Romano (1992): Bootstrap technology and applications. *Technometrics*, **34**, 378–398.

Livezey, R. E. (1985): Statistical analysis of general circulation model climate simulation: Sensitivity and prediction experiments. *J. Atmospheric Sci.*, **42**, 1139–1149.

Lukas, R., and E. Lindstrom (1991): The mixed layer of the western equatorial Pacific Ocean. *J. Geophys. Res.*, **96** (Supplement), 3343–3357.

Mason, S. J., and G. M. Mimmack (1992): The use of bootstrap confidence intervals for the correlation coefficient in climatology. *Theoret. Appl. Climatology*, **45**, 229–233.

McClave, J. T., and F. H. Dietrich, II (1991): "Statistics", Dellen (division of Macmillan Co., New York).

Mysak, L. A., and D. K. Manak (1989): Arctic sea-ice extent and anomalies, 1953–1984. *Atmosphere-Ocean*, **27**, 376–405.

Neave, H. R., and P. L. Worthington (1988): "Distribution-Free Tests", Unwin Hyman, London.

Pearson, E. S., and H. O. Hartley (1966): "Biometrika Tables for Statisticians", Cambridge Univ. Press, Cambridge, 264pp.

Preisendorfer, R. W., and C. D. Mobley (1988): "Principal Component Analysis in Meteorology and Oceangraphy", Elsevier, New York, 425pp.

Ramage, C. S. (1983): Teleconnections and the seige of time. *J. Climatology*, **3**, 223–231.

Ripley, B. (1987): "Stochastic Simulation", Wiley, New York, 237pp.

Ryan B. F., B. L. Joiner, and T. A. Ryan, Jr. (1992): "MINITAB Handbook, 2nd Ed.," PWS-KENT, Boston, 409pp.

Scheffé, H. (1959): "The Analysis of Variance", Wiley, New York, 477pp.

Shaw, W. J., R. L. Pauley, T. M. Gobel, and L. F. Radke (1991): A case study of atmospheric boundary layer mean structure for flow parallel to the ice edge: Aircraft observations from CEAREX. *J. Geophys. Res.*, **96**, 4691–4708.

Solow, A. R. (1985): Bootstrapping correlated data. *Math. Geo.*, **17**, 769–775.

Swanepoel, H. W. H., and J. W. J. van Wyk (1986): The bootstrap applied to spectral density function estimation. *Biometrika*, **73**, 135–142.

Thiébaux, H. J., and M. A. Pedder (1987): "Spatial Objective Analysis, with Applications in Atmospheric Science", Academic Press, London, 299pp.

Thiébaux, H. J., and F. W. Zwiers (1984): The interpretation and estimation of effective sample size. *J. Climate Appl. Meteorol.*, **23**, 800–811.

Wu, C. F. J. (1986): Jackknife, bootstrap and other resampling methods in regression analysis. *Ann. Statist.*, **14**, 1261–1295.

Zwiers, F. W. (1990): The effect of serial correlation on statistical inferences made with resampling procedures. *J. Climate*, **3**, 1452–1461.

Zwiers, F. W., and H. J. Thiébaux (1987): Statistical considerations for climate experiments. Part I: Scalar tests. *J. Climate Appl. Meteorol.*, **26**, 464–476.

Zwiers, F. W. (1987): Statistical considerations for climate experiments. Part II. Multivariate tests. *J. Climate Appl. Meteorol.*, **26**, 477–487.

INDEX